FORSCHUNGSBERICHTE AUS DEM LEHRSTUHL FÜR REGELUNGSSYSTEME

TECHNISCHE UNIVERSITÄT KAISERSLAUTERN

Band 19

T0133345

Forschungsberichte aus dem Lehrstuhl für Regelungssysteme

Technische Universität Kaiserslautern

Band 19

Herausgeber:

Prof. Dr. Steven Liu

Alen Turnwald

Modelling and Control
of an Autonomous Two-Wheeled Vehicle

Logos Verlag Berlin

 λογος

Forschungsberichte aus dem Lehrstuhl für Regelungssysteme
Technische Universität Kaiserslautern

Herausgegeben von
Univ.-Prof. Dr.-Ing. Steven Liu
Lehrstuhl für Regelungssysteme
Technische Universität Kaiserslautern
Erwin-Schrödinger-Str. 12/332
D-67663 Kaiserslautern
E-Mail: sliu@eit.uni-kl.de

Bibliographic information published by the Deutsche Nationalbibliothek

The Deutsche Nationalbibliothek lists this publication in the Deutsche
Nationalbibliografie; detailed bibliographic data are available
on the Internet at http://dnb.d-nb.de .

ISBN 978-3-8325-5205-3
ISSN 2190-7897

Logos Verlag Berlin GmbH
Georg-Knorr-Str. 4, Geb. 10,
12681 Berlin
Tel.:   +49 (0)30 / 42 85 10 90
Fax: +49 (0)30 / 42 85 10 92
http://www.logos-verlag.de

Modelling and Control of an Autonomous Two-Wheeled Vehicle

Modellierung und Regelung eines autonomen Zweiradfahrzeuges

Vom Fachbereich Elektrotechnik und Informationstechnik

der Technischen Universität Kaiserslautern

zur Verleihung des akademischen Grades

Doktor der Ingenieurwissenschaften (Dr.-Ing.)

genehmigte Dissertation

von

Alen Turnwald

geboren in Teheran

D 386

Tag der mündlichen Prüfung:	21.09.2020
Dekan des Fachbereichs:	Prof. Dr.-Ing. Ralph Urbansky
Vorsitzender der Prüfungskommission:	Prof. Dr. Stefan Götz
1. Berichterstatter:	Prof. Dr.-Ing. Steven Liu
2. Berichterstatter	Prof. Dr.-Ing. Paolo Mercorelli

Für Veronika
und Nora

Vorwort

Der Inhalt dieser Dissertation entstand während meiner Tätigkeit als wissenschaftlicher Mitarbeiter am Lehrstuhl für Regelungssysteme, Fachbereich Elektro- und Informationstechnik der technischen Universität Kaiserslautern.

An erster Stelle möchte ich Herrn Prof. Dr.-Ing. Steven Liu für seine Unterstützung bei der Erstellung dieser Arbeit danken. Er gab mir die Freiheit, an Themen meiner Wahl zu arbeiten, und förderte die Veröffentlichung meiner Ergebnisse. Die Diskussionen mit ihm regten mich stets zum Nachdenken an und haben meinen Blick auf Dinge nachhaltig verändert. Ich danke Herrn Prof. Dr.-Ing. Paolo Mercorelli für sein Interesse an meiner Arbeit und die Übernahme des Gutachtens. Mein Dank gilt auch Herrn Prof. Dr. Stefan Götz für den Vorsitz des Prüfungsausschusses.

Des Weiteren möchte ich meinen ehemaligen Kollegen danken, durch die ich die Zeit in Kaiserslautern als sehr angenehm empfand. Danke an Felix Berkel für die Begleitung auf mehreren Auslandsreisen. Danke an Sebastian Caba für ständige Empfehlung bester Literatur und an Markus Bell für nette Zugfahrten. Danke an Markus Lepper und Jawad Ismail für das Organisieren sportlicher Aktivitäten sowie an Fabian Kennel und Tim Steiner mit der bequemen Couch im Büro. Auch danke ich Herrn Prof. Daniel Görges für die wertvollen Diskussionen und Zusprüche. Weiterhin möchte ich Sanad Al-Areqi, Filipe Figueiredo, Yanhao He, Tim Nagel, Sven Reimann, Christian Tuttas, Yun Wan, Hengyi Wang, Benjamin Watkins, Min Wu, Yakun Zhou, Chirstoph Mark, Xiang Chen, Muhammad Ikhsan, Kashif Iqbal und Pedro Dos Santos für die Zusammenarbeit danken. Ein Dank geht ebenfalls an Jutta Lenhardt für die administrative Unterstützung. Zudem möchte ich Swen Becker für den vielseitigen und erweiterten Support während der Fertigstellung meiner Arbeit danken. Thomas Janz möchte ich an dieser Stelle besonders danken. Er stand mir während der praktischen Umsetzung stets mit Rat und Tat zur Seite. Ich habe sehr viel von ihm gelernt.

Außerdem möchte ich meinen *EFS*-Kollegen danken. Herr Dr. Stephan Neumaier hielt mir während der Fertigstellung der schriftlichen Ausarbeitung den Rücken frei. Weiter danke ich Herrn Dr. Christian Breindl für den konstruktiven fachlichen Austausch und Herrn Frieder Gottmann für die wertvollen Anregungen.

Zum Schluss möchte ich meiner Frau, Veronika, meinen besonderen Dank aussprechen, der ich diese Arbeit freudig widme. Sie begleitete mich unermüdlich während all dieser Zeit. Danke Dir für die Geduld, Hingabe und die Ermutigungen, insbesondere in den letzten Monaten meiner Tätigkeit an der Universität. Ohne Deine Unterstützung wäre diese Dissertation nicht entstanden.

Alen Turnwald   Ingolstadt den 09.10.2020

Contents

Notation

Tensors and their notation

Matrices and vectors are denoted by bold symbols such as $\boldsymbol{q}$. A tensor is denoted by a regular symbol such as q, and the elements of the tensor are denoted using sub- and superscripts such as q^i or p_α. If a symbol contains an index initially such as $M_\mathcal{D}$, the sub- and superscripts denoting the elements of the tensor are separated by a comma. For instance, $M_{\mathcal{D},ij}$ denotes the ij-th element of the tensor $M_\mathcal{D}$ or, equivalently, the ij-th entry of the matrix $\boldsymbol{M}_\mathcal{D}$. Calculating with tensors, Einstein's sum convention holds that is, for instance, $\boldsymbol{M}\,\boldsymbol{q} = M_{ij}\,q^j = \sum_{j=1}^{n} M_{ij}\,q^j$.

Mechanical and nonholonomic systems, Chapters 4

n	Number of the generalised coordinates
κ	Number of (nonholonomic) constraints

Following indexes are used for the tensors:

$i, j \in \{1, \cdots, n\}$	For generalised coordinates, e.g. q^i
$\alpha, \beta, \gamma, \mu \in \{1, \cdots, n - \kappa\}$	For base coordinate, e.g. r^α
$a, b, c \in \{1, \cdots, \kappa\}$	For fiber coordinates, e.g. s^a

$\mathcal{Q}$	Configuration space
$\mathcal{T}$	Tangent space
$\mathcal{T}^*$	Co-tangent space
$\mathcal{D}$	Reduced space on the tangent bundle $\mathcal{TQ}$
$\mathcal{M}$	Reduced space on the co-tangent bundle $\mathcal{T}^*\mathcal{Q}$
L	Lagrangian function or the Lagrangian
$L_\mathcal{D}$	Constrained Lagrangian function
H	Hamiltonian function or the Hamiltonian
$H_\mathcal{M}$	Constrained Lagrangian function
T	Kinetic energy
U	Potential energy
$\boldsymbol{q}$ or q^i	Generalised coordinates of $\mathcal{Q}$
$\boldsymbol{p}$ or p_j	Generalised impulse coordinates of $\mathcal{T}^*$

$\boldsymbol{r}$ or r^α	Coordinates of the horizontal spaces of $\mathcal{D}$ and $\mathcal{M}$, also base coordinates
$\boldsymbol{s}$ or s^a	Coordinates of the vertical spaces of $\mathcal{D}$ and $\mathcal{M}$, also fiber coordinates
$\boldsymbol{\rho}$ or ρ_α	Transformed impulse coordinates corresponding to the horizontal space of $\mathcal{M}$
$\breve{\boldsymbol{\rho}}$ or ρ_a	Transformed impulse coordinates corresponding to the vertical space of $\mathcal{M}$
$\boldsymbol{M}(\boldsymbol{q})$	Mass matrix
M_{ij}	Mass tensor, containing the elements of the mass matrix
$\boldsymbol{M}_\mathcal{D}(\boldsymbol{r})$	Reduced mass matrix
$M_{\mathcal{D},\alpha\beta}$	Reduced mass tensor, containing the elements of the reduced mass matrix
$\boldsymbol{M}^{-1}(\boldsymbol{q})$	Inverse mass matrix
M^{ij}	Inverse mass tensor, containing the elements of the inverse mass matrix
$\boldsymbol{M}_\mathcal{D}^{-1}(\boldsymbol{r})$	Inverse reduced mass matrix
$M^{\mathcal{D},\alpha\beta}$	Inverse reduced mass tensor, containing the elements of $\boldsymbol{M}_\mathcal{D}^{-1}$
A	Ehresmann connection
B	Curvature of the Ehresmann connection A
Γ	Tensor of the partial differential of $M_\mathcal{D,}$ w.r.t. r
T	Kinetic energy
U	Potential energy
T_v	Translational kinetic energy
T_Ω	Rotational kinetic energy

Sections 4.3 and 4.4

Lower-left indexes denote the Coordinate System (CS) a vector is given in. For instance, $_Z\boldsymbol{v}$ is a vector denoted by $\boldsymbol{v}$ given in the coordinate system Z. Rotation matrices are denoted by $\boldsymbol{R}$ where upper-left indexes specify the to-CS and from-CS accordingly. For instance, $^{SZ}\boldsymbol{R}$ transforms a vector given in Z into the corresponding vector given in S, in other words $_S\boldsymbol{v} = {}^{SZ}\boldsymbol{R}\,_Z\boldsymbol{v}$. Rotation of a CS with respect to another one is denoted by the angular velocity vector $\boldsymbol{\omega}$. For instance, the rotation of the CS B with respect to the CS I is $_Z^{IB}\boldsymbol{\omega}$, where the vector itself is given in the CS Z. By $Rot_1(\varphi)$ for instance, a standard rotation matrix about the first axis by an angle φ. For space saving, in this chapter we replace sin and cos by c and s, for example $s_\varphi := \sin\varphi$ and $c_\varphi := \cos\varphi$.

$\boldsymbol{w}$	A position vector
$\boldsymbol{v}$	A translational velocity vector
$\boldsymbol{\Omega}$	An angular velocity vector
x, y	Coordinates of the world coordinate system
ξ, η	Coordinates of the body-fixed coordinate system
δ	Real steering angle
$\tilde{\delta}$	Physical steering angle
ψ	Orientation angle
φ	Leaning angle
Δ	Trail
m	Total mass of the rear frame

m_s	Total mass of the steering assembly
l	Wheelbase
l_r, h	Centre of mass of the rear frame
d_s, h_s	Centre of mass of steering mechanism
J	Inertia of the rear frame
J_s	Inertial of the steering assembly
J_x, J	Components of the inertia of the rear frame
J_{sx}, J_s	Components of the inertia of the steering mechanism
J_w	Inertia of the wheel
R_w	Wheel radius
ϵ	Head angle
u_δ	Torque applied to the steering angle
u_φ	Torque applied to the leaning angle
u_ξ	Force applied on the rear wheel
σ	Steering variable
φ^*	Desired leaning angle
δ^*	Desired steering angle
θ	Angle of the body of rider

Optimisation and trajectory planning, Chapter 5

$\boldsymbol{x}$	State vector of a dynamical system
$\boldsymbol{u}$	Input vector of a dynamical system
$\boldsymbol{x}_0$	Initial state vector
$\boldsymbol{u}_0$	Initial input vector
$\boldsymbol{x}_F$	Final state vector
$\boldsymbol{u}_F$	Final input vector
t_0	Starting time
t_F	Final time
$\mathcal{C}$	The set corresponding to the constraints
$\mathcal{X}$	The set corresponding to the constraints on the states
$\mathcal{U}$	The set corresponding to the constraints on the inputs
$\tilde{\mathcal{J}}$	Objective of a continuous-time optimal control problem
$\mathcal{J}$	Objective of a nonlinear program
$\boldsymbol{z}$	Vector of decision variables
$\boldsymbol{b}_l$	Lower bound for $\boldsymbol{z}$
$\boldsymbol{b}_u$	Upper bound for $\boldsymbol{z}$
$\boldsymbol{x}_l$	Lower bound for $\boldsymbol{x}$
$\boldsymbol{x}_u$	Upper bound for $\boldsymbol{x}$
$\boldsymbol{u}_l$	Lower bound for $\boldsymbol{u}$
$\boldsymbol{u}_u$	Upper bound for $\boldsymbol{u}$
$\boldsymbol{X}$	Vector containing a path

$\boldsymbol{x}^\star$	Optimal state vector
$\boldsymbol{u}^\star$	Optimal input vector

Passivity-based control and port-Hamiltonian systems, Chapter 6

$\boldsymbol{J}(\boldsymbol{q},\boldsymbol{p})$	Interconnection matrix
$\boldsymbol{R}(\boldsymbol{q},\boldsymbol{p})$	Damping matrix
$\boldsymbol{J}_d(\boldsymbol{q},\boldsymbol{p})$	Desired interconnection matrix
$\boldsymbol{R}_d(\boldsymbol{q},\boldsymbol{p})$	Desired damping matrix
H_d	Desired Hamiltonian
$\boldsymbol{G}_{pH}$	Input matrix for a general port-Hamiltonian system
$\boldsymbol{G}_u$	Input matrix for mechanical port-Hamiltonian system with $\boldsymbol{G}_{\mathrm{pH}} = \begin{bmatrix} \mathbf{0} \\ \boldsymbol{G}_u \end{bmatrix}$
$\boldsymbol{y}_{\mathrm{passive}}$	Output of a passive system
$\boldsymbol{G}_u^\perp$	Annihilator to $\boldsymbol{G}_u$
$\Phi(\boldsymbol{x},t)$	A coordinate transformation
H_{add}	Additive Hamiltonian in a coordinate transformation
$\boldsymbol{\alpha}$	Additive output in a coordinate transformation
$\boldsymbol{\beta}$	Additive input in a coordinate transformation
$\boldsymbol{x}^\star$	Desired state vector
$\boldsymbol{u}^\star$	Desired input vector
$\boldsymbol{x}_I$	Vector of the integral states
$\boldsymbol{z}$	Vector of unknown parameter estimation errors
$u_{\varphi,\mathrm{dist}}$	Disturbance torque on the leaning angle
$*^p$	Desired values of a given path

Acronyms

TUK	Technische Universität Kaiserslautern
NLP	Nonlinear Program
LPV	Linear Parameter-Varying
CS	Coordinate System
CoM	Centre Of Mass
ODE	Ordinary Differential Equation
DAE	Differential-Algebraic Equation
PDE	Partial Differential Equation
TWV	Two-Wheeled Vehicle
EoM	Equation of Motion
CMG	Control Momentum Gyro
IDA	Iterconnection and Damping Assignment
PBC	Passivity-Based Controller
GCT	Generalised Canonical Transfomration
PH	Port-Hamiltonian
NIC	Network Interface Card
EtherCAT	Ethernet for Control Automation Technology
IMU	Inertial Measurement Unit
UCU	User Control Unit
PDMU	Power Distribution and Managemetn Unit
PU	Processing Unit
MC	Motion Controller
BLDC	Brushless Direct Current
GPS	Global Positioning System
SPI	Serial Peripheral Interface
SIP	System in Package
MEMS	Micro Electro Mechanical System
I2C	Inter-Integrated Circuit
NMEA	National Marine Electronics Association
PI	Proportional-Integral

1 Introduction

1.1 Motivation

With respect to the future urban mobility, modern electrical bicycles, advanced motorcycles and innovative two-wheeled vehicles are arresting enormous amount of attention. The dynamic behaviour of two-wheeled vehicles, especially their self-stability, has been matter of scientific studies for many decades. In fact, as stated in [MP11], the self-stability of bicycles as dynamical systems have been studied as early as in 1910. The necessity of serious investigations of two-wheeled vehicles from a control engineering perspective, however, has increased in the last several years. Comparing to four-wheeled vehicles, an autonomous driving bicycle for passenger transportations may today sound like science-fiction or even unnecessary. Yet, current evolution of the technology towards intelligent, automated and highly connected mobility of the future requires the machine intelligence to be able to handle two-wheeled vehicles as well.

The advantages of two-wheeled vehicles with respect to energy consumption, parking space and environmental friendliness are self-evident. An autonomous two-wheeled vehicle is for instance, among many other conceivable application fields, a noticeable candidate for future automated delivery systems in urban environments. Therefore, model-based control and optimal trajectory planning for such vehicles will remain inevitable matter of research and development in the future. The results are currently, and will in short term be further used for developing safety-increasing assistant systems for existing vehicles, e.g. [KMT09]. In long term, developing innovative autonomous two-wheeled vehicles[1] is likely to require even more intensive research, as it is the case for autonomous driving cars today.

While a large portion of the available know-how on sensor data processing, localisation methods and similar technologies are transferable from cars, due to the special dynamical behaviour of two-wheeled vehicles, methods for trajectory planning and tracking control are not trivially reusable. Therefore, a reliable and yet usable vehicle model as well as a systematic approach to motion control for two-wheeled vehicles are essential, to which the present work makes a contribution.

---

[1]For instance C1 from Lit Motors: `www.litmotors.com/product`

1.2 Objective and contribution of this work

The available literature contains, on the one hand, extensive studies of the dynamics of bicycles based on quantitative considerations and numeric calculations, which lead to sophisticated models that are not trivially usable for systematic control synthesis. On the other hand, simpler bicycle models are used for purposes of trajectory planing and motion control synthesis. These models however, have either a constrained range of validity, such as models for constant velocities, or, ignore parts of bicycle's physics to keep the resulting equations simple. The main objective of this thesis is to fill this gap and to present a unifying approach to modelling and control for autonomous two-wheeled vehicles. The resulting model shall be generally valid and physically detailed enough to represent the characteristic dynamical behaviour, and at the same time, be proper for a systematic control synthesis. Furthermore, an extension of the model by a rider or further actuators is required to be possible in a systematic way.

To this end, the Hamiltonian framework is chosen. The main contribution of this work is to propose a vehicle model as a port-Hamiltonian system, which is derived using an automatable scheme. The model represents the bicycle's physics such as the self-stability and is, at the same time, directly usable for model-based trajectory planing as well as for passivity-based trajectory tracking controller design. The proposed methodic approach to model derivation is a general procedure that can be extended in a systematic way.

Furthermore, a trajectory tracking controller is designed that is physically interpretable, is valid for the sophisticated vehicle model, and, also robust against parameter uncertainties. To this end, existing approaches of passivity-based controller design are extended and adjusted for two-wheeled vehicles.

The outline of this thesis and particular contributions in every chapter are given below.

Chapter 3
In this chapter, the physics of bicycles is summarised and facts about their dynamics are explained, which are recalled throughout the entire work. Furthermore, the so-called linear benchmark model is presented which is taken as a basis to validate dynamical behaviour of the vehicle model proposed in this thesis.

Chapter 4
In this chapter, the main contribution of this work is presented. That is proposing an approach to derive a physically realistic, yet usable model for two-wheeled vehicles for control synthesis, as well as for trajectory planning. Some geometric considerations at the beginning of the model derivation make sure that physical phenomenon such as self-stability is represented by the resulting model, that is validated later in the chapter. The model is validated by comparison to the benchmark model from the literature, as well as by a series of simulation scenarios, which are designed to demonstrate different aspects of

the dynamical behaviour of the vehicle. Further, a methodical approach is proposed for structure preserving simplification of the equations of motion as well as extension by a new rigid body, for instance an active rider. A part of the content of this chapter is published in [TL18b] and [TL18a].

Chapter 5
In this chapter, optimal trajectory planning methods are used to demonstrate the usability of the proposed model as well as the introduced simplification. It is shown that using the systematic model simplification, trajectories can be planned for the vehicle with a lower computation effort without significant loss of model accuracy. A part of the content of this chapter is published in [TL19].

Chapter 6
In this chapter, a passivity-based trajectory tracking controller is developed using the proposed vehicle model. To this end, existing approaches are extended and various simulations are run to demonstrate the performance of the closed loop under different conditions. Some initial ideas used in this chapter were published in [TO17]. A part of the content of this chapter is, furthermore, published in [TSL18] and [TL18a].

Chapter 7
In this chapter, a prototype two-wheeled vehicle is briefly introduced, which was developed related to this work.

Chapter 8
In this chapter experimental results are presented which demonstrate the functionality of the prototype vehicle and the concept of motion control for two-wheeled vehicles. Furthermore, experimental results are presented for the validation of the proposed model, as well as for the demonstration of its advantage comparing to a widely used nonlinear model from the literature. A part of the content of this chapter is published in [TL19].

1.3 Published content

Related to this work, following papers were published:

[TO17]   A. Turnwald and T. Oehlschlägel. Passivity-based control of a cryogenic upper stage to minimize fuel sloshing. *Journal of Guidance, Control, and Dynamics*, 2017.

[TL18b]   A. Turnwald and S. Liu. A nonlinear bike model for purposes of controller and observer design. *IFAC-PapersOnLine*, 2018. 9th Vienna International Conference on Mathematical Modelling.

[TL18a]   A. Turnwald and S. Liu. Adaptive trajectory tracking for a planar two-wheeled vehicle with positive trail. *In 2018 IEEE Conference on Control Technology and Applications (CCTA)*, 2018.

[TSL18]   A. Turnwald, M. Schäfer, and S. Liu. Passivity-based trajectory tracking control for an autonomous bicycle. *In IECON 2018 - 44th Annual Conference of the IEEE Industrial Electronics Society*, 2018.

[TL19]   A. Turnwald and S. Liu. Motion planning and experimental validation for an autonomous bicycle. *In IECON 2019 - 45th Annual Conference of the IEEE Industrial Electronics Society*, 2019.

Furthermore, some of the supervised masters theses are listed as:

Adalbert, M. (2016)	Control of port-Hamiltonian systems via generalised canonical transformations
Matheis, N. (2016)	Implementation of a passivity-based controller for an inverted rotational pendulum
Garcia, F. J. R. (2017)	Development and implementation of pose estimation and localization for an autonomous bicycle
Schäfer, M. (2018)	Passivity-based trajectory tracking control for a two-wheeled vehicle
Ahmed, R. (2019)	Model validation for a two-wheeled vehicle using multibody simulation and experimental data
Mouaffo, U. (2019)	Optimal path and trajectory planning for a two-wheeled vehicle using nonlinear dynamics
Muniappan, K. (2019)	Implementation and experimental validation of linear controllers for an autonomous bicycle
Thirumurugan, D. (2019)	Design and experimental investigation of a CMG-based stabilisation of a two-wheeled vehicle

2 Preliminaries

This chapter summarises some of the required definitions and the notation used in this work. Note that detailed explanations, derivations or proofs are omitted on purpose, since those are found in the literature cited within the text of this chapter.

In Section 2.1, the tensor notation, that is mainly used in Chapter 4, is briefly described. Further, some necessary preliminaries from the systems theory are mentioned, especially for Chapters 4 and 6. Finally, some basic definitions are given in context of mechanical systems. In Section 2.2, modelling of constrained mechanical systems is addressed and the derivation of some essential equations from the literature is presented.

2.1 General notations and elementary definitions

2.1.1 Tensor notation

For sake of clarity and comprehension, the tensor notation used in this work, mainly in Chapter 4, is outlined briefly. Tensor notation is a mathematical tool from tensor calculus that is often used in different fields of physics, especially relativity theory, since it allows generic handling of multi-dimensional entities. The following introduction to tensor notation is mainly based on [Sus10] and [Bis16].

Within the present work, entities with more than one components, in other words non-scalars, are denoted using matrix notation or tensor notation in an equivalent meaning. Note that in this thesis, the definitions of vectors or tensors are not considered strictly, rather only the notations and mathematical tools from the corresponding calculus are applied for consistent calculations. Matrices and vectors are denoted by bold symbols, for instance the vector

$$\boldsymbol{f} \in \mathbb{R}^3 \, .$$

Using the the tensor notation, an entity is denoted by indexes as sub- and/or superscripts. For example

$$f_\alpha \text{ with } \alpha = 1, 2, 3.$$

Note that it is only a matter of convention whether an index is a sub- or a superscript. This means that the same vector can also be denoted as f^α, as long as this *choice* is consistent throughout the entire calculations.

The number of indexes determines the dimension of the entity. For instance the matrix $\boldsymbol{M}$ can be equivalently denoted by

$$M_{\alpha\beta} := \boldsymbol{M} \in \mathbb{R}^{n\times m}$$
$$\text{or } M^{\beta}_{\alpha} := \boldsymbol{M} \in \mathbb{R}^{n\times m}$$
$$\text{or } M^{\alpha}_{\beta} := \boldsymbol{M} \in \mathbb{R}^{n\times m}$$
$$\text{or } M^{\alpha\beta} := \boldsymbol{M} \in \mathbb{R}^{n\times m},$$

with $\alpha = 1, \cdots, n$ and $\beta = 1, \cdots, m$. Note that for every combination (α, β), the above symbols denote the $\alpha\beta$-component of the matrix $\boldsymbol{M}$.

In tensor calculus, components corresponding to the subscripts are referred to as *co-variants* and those corresponding to superscripts as *contra-variant*. This has to do with the way how tensors are transformed with regard to their basis. In Cartesian coordinates, the co- and contra-variants are identical. In this work, the sub- and superscripts are used to consistently denote specific entities such as generalized coordinates and impulses in accordance to the notation from the book [Blo16].

Once the convention is chosen, every denotation must be consistent and the order of the indexes is substantial. For instance, if the notation $M_{\alpha\beta} := \boldsymbol{M}$ is chosen, the transposed matrix is denoted in tensor form as $M_{\beta\alpha} := \boldsymbol{M}^{T}$.

The most important advantage of the tensor notation, and the main reason for using them in this work, is the fact that entities with a dimension > 2 can be denoted simply and generically. An example is the three-dimensional tensor

$$B^{b}_{\alpha\beta} \qquad\qquad \text{with} \qquad\qquad \alpha = \beta = 1 - n_r \text{ und } b = 1 - n_s,$$

which will be defined in Section 2.2. Note that this entity cannot be denoted and calculated with using matrix notation only.

Using the tensor notation, the so-called *Einstein's sum convention* holds. That is, when summing over a particular index, where the index is a subscript in one tensor and a superscript in the other, the sum-symbol is omitted. For instance, the scalar product of two vectors $\boldsymbol{f}$ and $\boldsymbol{g}$

$$\boldsymbol{f}^{T}\boldsymbol{g} = f_1\,g_1 + f_2\,g_2 + \cdots + f_n\,g_n = \sum_{i=1}^{n} f_i\,g_i$$

is given using tensor notation as

$$\boldsymbol{f}^{T}\boldsymbol{g} = f_i\,g^{i} = f^{i}\,g_i = g_i\,f^{i} = g^{i}\,f_i$$

depending on the chosen convention. Note that since the product $g^{i}\,f_i$ is the product of two scalars, namely the ith component of each vector, the order can be changed.

When the sum convention is applied, the corresponding index vanishes in the product result. For instance

$$p_i = M_{ij}\, \dot{q}^j = M_{ik}\, \dot{q}^k \ .$$

Note that the index over which the summation is applied can be arbitrarily exchanged $(j \rightarrow k)$ as long as it is the same index for both tensors. In accordance to the most literature involving tensor notation, often indexes are in fact reused in this work, over which the summation applies, to reduce the number of used letters.

Two tensors can only be added if they have the same indexes in the same order, for example

$$f^\beta_\alpha + g^\beta_\alpha = h^\beta_\alpha \quad \text{or} \quad M_{\alpha\beta} + N_{\alpha\beta} = P_{\alpha\beta}.$$

If a tensor is in the denominator of a fraction is chosen to have its indexes as subscript, it may be given in the nominator exchanging the subscript with the superscript, e.g.

$$f_\alpha = \frac{1}{f^\alpha} \quad \text{or} \quad f^\alpha = \frac{1}{f_\alpha}. \tag{2.1}$$

Therefore,

$$f^\beta_\alpha = g^\beta_\alpha + \frac{1}{h^\alpha_\beta} + \frac{m_\alpha}{n_\beta}. \tag{2.2}$$

is a valid equation since the indexes are consistent.

For more details on tensor calculus and the tensor notation, one may refer to the lectures of Prof. Leonard Susskind ([Sus10]) and also in [DP10].

2.1.2 System theory

A general nonlinear time-invariant system $\sum$ is given by

$$\sum : \begin{cases} \dot{\boldsymbol{x}} &= \boldsymbol{f}(\boldsymbol{x}, \boldsymbol{u}) \\ \boldsymbol{y} &= \boldsymbol{h}(\boldsymbol{x}, \boldsymbol{u}) \end{cases} , \boldsymbol{x} \in \mathcal{X}\,, \boldsymbol{y} \in \mathcal{Y}\,, \boldsymbol{U} \in \mathcal{U} \tag{2.3}$$

where $\boldsymbol{u}(t)$ is the input vector, $\boldsymbol{y}(t)$ the output vector and the manifolds $\mathcal{U}$ and $\mathcal{Y}$ are the corresponding domains. $\boldsymbol{x} = [x_1, x_2, ...x_n]^T$ is the state vector and the manifold $\mathcal{X}$ is the corresponding domain.

Definition 2.1.1.
The system $\sum$ is called autonomous, if $\boldsymbol{u}(t) \equiv \boldsymbol{0}$.

Definition 2.1.2. *[Ada15]*
A system is called **input-affine***, if it is defined as*

$$\sum_{aff} : \begin{cases} \dot{\boldsymbol{x}} = \boldsymbol{f}(\boldsymbol{x}) + \boldsymbol{g}(\boldsymbol{x})\boldsymbol{u} \\ \boldsymbol{y} = \boldsymbol{h}(\boldsymbol{x}) \end{cases} . \tag{2.4}$$

Definition 2.1.3. *[FS01a]*
The system $\sum_{aff}$ *is called* **distinguishable***, if for a given trajectory* $\boldsymbol{x}_d(t)$ *satisfying* (2.4)

$$\boldsymbol{x}(t_0) = \boldsymbol{x}_d(t_0) \; , \; \boldsymbol{y} - \boldsymbol{y}_d \equiv \boldsymbol{0} \; \forall t \in [t_0, t_1] \;\; \Rightarrow \;\; \boldsymbol{x}(t) = \boldsymbol{x}_d(t) \; , \forall t \in [t_0, t_1] \tag{2.5}$$

holds.

Definition 2.1.4.
The point $\boldsymbol{x}^*$ *is called an* **equilibrium** *for the system* $\sum$ *if*

$$\dot{\boldsymbol{x}}^* = \boldsymbol{f}(\boldsymbol{x}^*) = \boldsymbol{0} \; . \tag{2.6}$$

Definition 2.1.5. *[Ada15]*
The equilibrium $\boldsymbol{x}^*$ *of the autonomous system* $\sum$ *is said to be* **locally attractive***, if a neighbourhood* $U(\boldsymbol{x}^*) \subseteq \mathcal{X}$ *exists, such that every initial value*

$$\boldsymbol{x}_0 \in U(\boldsymbol{x}^*) \tag{2.7}$$

leads to a trajectory $\boldsymbol{x}(t)$ *converging to the equilibrium for* $t \to \infty$*. If the neighbourhood is the entire space* $U(\boldsymbol{x}^*) = \mathcal{X}$ *the equilibrium is said to be* **globally attractive***.*

Definition 2.1.6. *[Lun16]*
The equilibrium $\boldsymbol{x}^* = \boldsymbol{0}$ *of the autonomous system* $\sum$ *is said to be* **stable** *according to* **Lyapunov***, if for every* ϵ*, a* δ *exists such that:*

$$\|\boldsymbol{x}_0\| < \delta(\epsilon) \;\; \Rightarrow \;\; \|\boldsymbol{x}(t)\| < \epsilon \; \forall t > 0 \; . \tag{2.8}$$

Definition 2.1.7. *[Ada15] [Lun16]*
The equilibrium $\boldsymbol{x}^*$ *is said to be locally (globally)* **asymptotically stable** *if it is stable according to Lyapunov and, furthermore, locally (globally) attractive.*
Or: if $\boldsymbol{x}^*$ *f it is stable according to Lyapunov and, furthermore*

$$\lim_{t \to \infty} \|\boldsymbol{x}(t)\| = \boldsymbol{x}^* \tag{2.9}$$

holds.

Definition 2.1.8. *[Ada15], [Lun16]*
The function $V(\boldsymbol{x}) : U(\boldsymbol{x}^*) \to \mathbb{R}$ *is called a* **Lyapunov-function** *for the autonomous system* $\sum$*, if it fulfils the following conditions:*

1. $V(\boldsymbol{x})$ *is continuous,* $V(\boldsymbol{x} = \boldsymbol{x}^*) = 0$ *and* $\frac{\partial V(\boldsymbol{x})}{\partial \boldsymbol{x}}$ *exists.*

2. $V(\boldsymbol{x}) > 0$ *for* $\boldsymbol{x} \neq \boldsymbol{x}^*$

3. $\dot{V}(\boldsymbol{x}) = \dfrac{\partial^T V(\boldsymbol{x})}{\partial \boldsymbol{x}} \boldsymbol{f}(\boldsymbol{x}) \leq 0$

Definition 2.1.9. *The function $V(x)$ is called **radially unbounded** if*

$$\|x\| \to \infty \Rightarrow V(x) \to \infty \tag{2.10}$$

holds.

Following definitions are mainly based on [Sch00] and [Kot10].

Definition 2.1.10.
*The **inner product** of two signals $f(t)$ and $g(t)$ is defined by*

$$\langle f, g \rangle = \int_0^\infty f^T g \, dt \; . \tag{2.11}$$

Definition 2.1.11.
*The L_2-**norm** of a signal $f(t)$ is defined by*

$$\|f\|_2 = \sqrt{\langle f, f \rangle} \; . \tag{2.12}$$

Definition 2.1.12.
*A function $v(u, y) : \mathcal{U} \times \mathcal{Y} \to \mathbb{R}$ is called a **supply rate**.*

Definition 2.1.13.
*A function $S(x) : \mathcal{X} \to \mathbb{R}_+$ is called a **storage function**.*

Definition 2.1.14.
*A state space system $\sum$ is said to be **dissipative** with respect to the supply rate v, if there exists a storage function, such that for all $x_0 \in \mathcal{X}$, all $t_1 \geq t_0$, and all input functions $u(t)$*

$$S(x(t_1)) \leq S(x(t_0)) + \int_{t_0}^{t_1} v(u(t), y(t)) dt \tag{2.13}$$

where $x_0 = x(t_0)$.

Definition 2.1.14 states that the stored energy in the system at any future time t_1, $S(x(t_1))$, is *at most* equal to the stored energy at the present t_0 plus the externally supplied energy during the time interval $[t_0, t_1]$. In other words, without an external energy supplement, the stored energy in a system cannot increase, or the system can only *dissipate* energy and not create some.

Assuming differentiability of the storage function $S(x)$, the *dissipation inequality* (2.1.14) can also be given in differential form as

$$\dot{S}(x(t)) \leq v(u(t), y(t)) \; . \tag{2.14}$$

Definition 2.1.15.
*A system $\sum$ is called **lossless**, if in (2.13) or (2.14), equality holds.*

Definition 2.1.16. *Passivity*
A system $\sum$ is called

- *passive if it is dissipative according to Definition 2.1.14 with supply rate*

$$v(\boldsymbol{u}(t), \boldsymbol{y}(t)) = \langle \boldsymbol{y}(t), \boldsymbol{u}(t) \rangle = \boldsymbol{y}(t)^T \cdot \boldsymbol{u}(t) \ . \tag{2.15}$$

- *strictly input-passive if it is dissipative according to Definition 2.1.14 with supply rate*

$$v(\boldsymbol{u}(t), \boldsymbol{y}(t)) = \langle \boldsymbol{y}(t), \boldsymbol{u}(t) \rangle - \alpha \|u\|^2, \ \alpha > 0 \ . \tag{2.16}$$

- *strictly output-passive if it is dissipative according to Definition 2.1.14 with the supply rate*

$$v(\boldsymbol{u}(t), \boldsymbol{y}(t)) = \langle \boldsymbol{y}(t), \boldsymbol{u}(t) \rangle - \beta \|y\|^2, \ \beta > 0 \ . \tag{2.17}$$

Definition 2.1.17.
*A lossless passive system $\sum$ is called **conservative**.*

The supply rate from (2.15) is the inner product of the input and the output which may be interpreted as *power*. Two well-known examples are mechanical system with generalised forces as input and generalised velocities as output, and, electrical systems with voltages as input and the corresponding currents as output. In this sense, the differential passivity inequality (2.14) states that the increase rate of the energy in a system is bounded by the the power put into it.

Definition 2.1.18.
Suppose a function $f(\boldsymbol{x}) : \mathcal{X} \to \mathbb{R}$. A point $\boldsymbol{x}^* \in \mathcal{X}$ is called a

- *local minimum if $\exists \epsilon > 0 : f(\boldsymbol{x}^*) \le f(\boldsymbol{x}), \ \forall \boldsymbol{x}, \|\boldsymbol{x}\| < \epsilon$.*

- *global minimum if $f(\boldsymbol{x}^*) \le f(\boldsymbol{x}), \ \forall \boldsymbol{x} \in \mathcal{X}$.*

- *global (local) strict minimum if $<$ holds instead of $\le$.*

Theorem 2.1.19. *Stability of dissipative systems [Sch00]*
Suppose a system $\sum$ with an equilibrium $\boldsymbol{x}^*$ is dissipative with regard to the supply rate v according to Definition 2.1.14. Suppose further $\boldsymbol{u} \equiv \boldsymbol{0}$ and

$$v(\boldsymbol{0}, \boldsymbol{y}) \le 0, \ \forall \boldsymbol{y}. \tag{2.18}$$

$\boldsymbol{x}^*$ is locally stable acceding to Lyapunov if it is a local strict minimum of the storage function $S(\boldsymbol{x})$. Furthermore,

$$V(\boldsymbol{x}) = S(\boldsymbol{x}) . \tag{2.19}$$

is a Lyapunov-function.

Sine passive systems are an special case of dissipative systems the stability of passive systems is defined based on the stability of the general class, namely dissipative systems. For a passive system, the supply rate is defined as the inner product of the input and the output $\langle y(t), u(t) \rangle$ and, thus, condition (2.18) is directly satisfied. Therefore, the following can be stated:

Theorem 2.1.20. *Stability of passive systems*
An equilibrium x^* of a passive system is stable according to Lyapunov with the Lyapunov-function $V(x) = S(x)$ if it is a strict minimum of $S(x)$.

In other words, once a system is passive the strict minimum of its storage function is a stable equilibrium.

Theorem 2.1.21. *Asymptotic stability of passive system*
Given an input-affine system $\sum_{aff}$ that is both fully reachable and stabilisable. Suppose further that $\sum_{aff}$ is strictly output-passive with the storage function $S(x)$ such that

$$S(x) > 0, \; \forall x \neq 0, \quad S(0) = 0 . \tag{2.20}$$

Then $x^* = 0$ is a local asymptotic stable equilibrium according to Lyapunov.

Definition 2.1.22. *Generalised Hamiltonian systems [Sch00] [vdSJ14]*
A generalised Hamiltonian system is given by

$$\sum_{GpH} : \begin{cases} \dot{x} = (J(x) - R(x)) \frac{\partial^T H(x)}{\partial x} + G_{pH}(x)\, u, & x \in \mathcal{X}, \; u \in \mathcal{U} \\ y = G_{pH}^T(x) \frac{\partial^T H(x)}{\partial x} & y \in \mathcal{Y}. \end{cases} \tag{2.21}$$

*$H(x) : \mathcal{X} \to \mathbb{R}$ is called the **Hamiltonian** or the Hamiltonian function and the skew-symmetric matrix $J(x) \in \mathbb{R}^{n \times n}$*

$$J(x) = -J^T(x) \tag{2.22}$$

the structure matrix. The symmetric and positive semi-definite matrix $R(x) \in \mathbb{R}^{n \times n}$

$$R(x) = R^T(x) \geq 0. \tag{2.23}$$

is called the damping matrix.

$\sum_{GpH}$ is often called a **port-Hamiltonian** system since energy can be inserted into the system by the *port* containing the input and output (u, y). The structure matrix corresponds to the energy exchange in the system and the damping matrix corresponds to the energy dissipation. A Hamiltonian system with $R(x) = 0$ is a lossless system.

2.1.3 Mechanical systems

The class of mechanical systems is quite important to control theory. Also, since the systems considered throughout this work are all from this class, specific definitions with regard to mechanical systems are explained separately in the following sections.

A general unconstrained mechanical system is given by the equations of motion as

$$M(q)\ddot{q} + h(q, \dot{q}) = G u, \tag{2.24}$$

where $q = [q_1 \ \dots \ q_n]^T \in \mathcal{Q}$ are the **generalised coordinates** and the manifold $\mathcal{Q}$ is the **configuration space**. The symmetric positive definite matrix $M(q)$ is called the **mass matrix** and G is the **input matrix**. Furthermore, u are the **generalised forces** which act externally on the system.

Definition 2.1.23.
A mechanical system as in (2.24) is called

- *fully actuated if G has full rank, or $rank(G) = n$.*

- *under-actuated if $rank(G) < n$.*

Euler-Lagrange systems

A commonly used approach to derive equations of motion for mechanical systems is the so-called Euler-Lagrange method ([Sch00]). For that, first the kinetic energy $T(q, \dot{q})$ and the potential energy $U(q)$ of the system are determined in relation to the generalised coordinates and their derivatives. The so-called **Lagrangian** L is then calculated as

$$L(q, \dot{q}) = T - U. \tag{2.25}$$

Given the Lagrangian, the Euler-Lagrange-Equation

$$\frac{d}{dt}\left(\frac{\partial^T L(q, \dot{q})}{\partial \dot{q}}\right) - \frac{\partial^T L(q, \dot{q})}{\partial q} = \tau \tag{2.26}$$

is used to determine the equations of motion in which τ are the generalised non-conservative forces such as friction or external forces or torques.

Considering the structure of the kinetic energy in a mechanical system

$$T = \frac{1}{2}\dot{q}^T M(q)\dot{q}, \tag{2.27}$$

the partial derivative of the Lagrangian with respect to $\dot{q}$ is given by

$$\frac{\partial^T L(q, \dot{q})}{\partial \dot{q}} = M(q)\dot{q}. \tag{2.28}$$

Thus, the Euler-Lagrange equation can be rewritten as

$$\frac{d}{dt}\left(\boldsymbol{M}(\boldsymbol{q})\dot{\boldsymbol{q}}\right) = \frac{\partial^T L(\boldsymbol{q},\dot{\boldsymbol{q}})}{\partial \boldsymbol{q}} + \boldsymbol{\tau} \ . \tag{2.29}$$

Using the chain rule, (2.29) can be further rewritten as

$$\boldsymbol{M}(\boldsymbol{q})\ddot{\boldsymbol{q}} = -\frac{d}{dt}\left(\boldsymbol{M}(\boldsymbol{q})\right)\dot{\boldsymbol{q}} + \frac{\partial^T L(\boldsymbol{q},\dot{\boldsymbol{q}})}{\partial \boldsymbol{q}} + \boldsymbol{\tau} \ . \tag{2.30}$$

Equation (2.30) can be illustrated schematically as in Figure 2.1. Note that this diagram can directly be used for simulation of mechanical systems, for instance using Matlab/Simulink® .

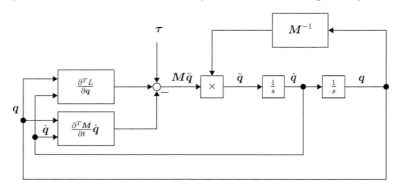

Figure 2.1: Structure of a Euler-Lagrange Systems according to (2.30)

Port-Hamiltonian mechanical systems

In what follows, a port-Hamiltonian system is a port-Hamiltonian mechanical system, if not explicitly denoted otherwise.

Definition 2.1.24.
Generalised impulses $\boldsymbol{p}$ are defined as the partial derivative of the Lagrangian with regard to the time-derivative of the generalised coordinates as

$$\boldsymbol{p} := \frac{\partial^T L(\boldsymbol{q},\dot{\boldsymbol{q}})}{\partial \dot{\boldsymbol{q}}} - \boldsymbol{M}(\boldsymbol{q})\dot{\boldsymbol{q}} \tag{2.31}$$

Considering from (2.31) that

$$\dot{\boldsymbol{q}} = \boldsymbol{M}^{-1}(\boldsymbol{q})\boldsymbol{p} \ , \tag{2.32}$$

the kinetic energy of a mechanical system can be formulated in terms of generalised impulses as

$$T(\boldsymbol{q},\boldsymbol{p}) = \frac{1}{2}\boldsymbol{p}^T \boldsymbol{M}^{-1}(\boldsymbol{q})\boldsymbol{p} \ . \tag{2.33}$$

Definition 2.1.25. *[Van05] [Ada15]*
*For a mechanical system, the **Hamiltonian** is defined as the sum of the kinetic and potential energy, that is*

$$H(\boldsymbol{q}, \boldsymbol{p}) = T(\boldsymbol{q}, \boldsymbol{p}) + U(\boldsymbol{q}) \tag{2.34}$$

$$H(\boldsymbol{q}, \boldsymbol{p}) = \frac{1}{2}\boldsymbol{p}^T \boldsymbol{M}^{-1}(\boldsymbol{q})\boldsymbol{p} + U(\boldsymbol{q}) \ . \tag{2.35}$$

Note that the Hamiltonian is the overall energy of a system and the potential energy is, in general, assumed to be bounded from below.

Definition 2.1.26. *Mechanical **port-Hamiltonian systems** [Van05] [vdSJ14]*
A lossless port-Hamiltonian system is given by

$$\sum_{pH} : \begin{cases} \dot{\boldsymbol{q}} = \dfrac{\partial^T H(\boldsymbol{q}, \boldsymbol{p})}{\partial \boldsymbol{p}}, & (\boldsymbol{q}, \boldsymbol{p}) = [q_1, ..., q_n, p_1, ..., p_n] \\[2ex] \dot{\boldsymbol{p}} = -\dfrac{\partial^T H(\boldsymbol{q}, \boldsymbol{p})}{\partial \boldsymbol{q}} + \boldsymbol{G}_u(\boldsymbol{q})\,\boldsymbol{u}, & \boldsymbol{u} \in \mathbb{R}^m \\[2ex] \boldsymbol{y} = \boldsymbol{G}_u(\boldsymbol{q})^T \dfrac{\partial^T H}{\partial \boldsymbol{p}}, & \boldsymbol{y} \in \mathbb{R}^m \end{cases} \tag{2.36}$$

Theorem 2.1.27. *Passivity of port-Hamiltonian systems*
A port-Hamiltonian system $\sum_{pH}$ with a positive definite Hamiltonian $H(\boldsymbol{q}, \boldsymbol{p}) > 0$ is passive.

The assumption of a positive definite Hamiltonian is not a significant restriction and simply justified for mechanical systems. The kinetic energy T is naturally positive definite due to its quadratic from with the positive definite mass matrix $\boldsymbol{M}$. One may interpret this fact by the following statement: If there is a motion in the system ($\dot{\boldsymbol{q}} \neq \boldsymbol{0}$) the kinetic energy is always positive, and otherwise it is zero. The potential energy is bounded from below, however, the lower bound of the potential energy varies depending on the choice of the origin of the considered coordinates.

Choosing the state vector $\boldsymbol{x}^T = \begin{bmatrix} \boldsymbol{q} & \boldsymbol{p} \end{bmatrix}$, the lossless port-Hamiltonian system (2.36) can be given in the generalised form as in (2.21) by

$$\sum_{pH} : \begin{cases} \begin{bmatrix} \dot{\boldsymbol{q}} \\ \dot{\boldsymbol{p}} \end{bmatrix} = \underbrace{\begin{bmatrix} \boldsymbol{0} & \boldsymbol{I} \\ -\boldsymbol{I} & \boldsymbol{0} \end{bmatrix}}_{\boldsymbol{J}} \begin{bmatrix} \frac{\partial^T H(\boldsymbol{q}, \boldsymbol{p})}{\partial \boldsymbol{q}} \\ \frac{\partial^T H(\boldsymbol{q}, \boldsymbol{p})}{\partial \boldsymbol{p}} \end{bmatrix} + \underbrace{\begin{bmatrix} \boldsymbol{0} \\ \boldsymbol{G}_u(\boldsymbol{q}) \end{bmatrix}}_{\boldsymbol{G}_{\mathrm{pH}}(\boldsymbol{x})} \boldsymbol{u} \\[4ex] \boldsymbol{y} = \begin{bmatrix} \boldsymbol{0} & \boldsymbol{G}_u^T(\boldsymbol{q}) \end{bmatrix} \begin{bmatrix} \frac{\partial^T H(\boldsymbol{q}, \boldsymbol{p})}{\partial \boldsymbol{q}} \\ \frac{\partial^T H(\boldsymbol{q}, \boldsymbol{p})}{\partial \boldsymbol{p}} \end{bmatrix} \end{cases} \tag{2.37}$$

For the illustration and simulation of a port-Hamiltonian system, one may directly use (2.37) which leads to the diagram given in Figure 2.2.

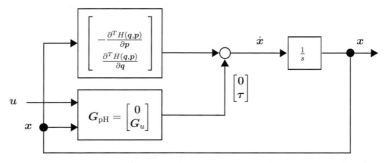

Figure 2.2: Structure of a port-Hamiltonian system according to (2.37)

It is worth mentioning that the definition of the impulse coordinates p can also help for the simulation and illustration of Euler-Lagrange systems. Considering

$$\frac{d\left(M\dot{q}\right)}{dt} = \frac{dp}{dt} , \qquad (2.38)$$

the equation (2.29) can be used as shown in Figure 2.3. Note that in this figure the impulse coordinates are stated as function $p\left(q,\,\dot{q}\right)$.

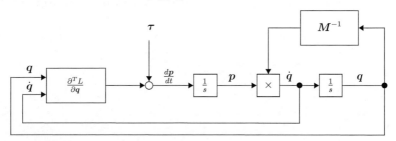

Figure 2.3: Structure of a Euler-Lagrange system according to (2.29)

Stability of port-Hamiltonian systems

The stability investigation for port-Hamiltonian systems is often based on their natural passivity.

Theorem 2.1.28. *Stability of port-Hamiltonian systems [Sch00]*
Let $\sum_{pH}$ be a port-Hamiltonian system with the potential energy $U(q)$. If q^* is a strict

minimum of the potential energy $U(\boldsymbol{q})$, then $\boldsymbol{x}^* = [\boldsymbol{q}^*\ 0]^T$ is a locally stable equilibrium according to Lyapunov with the Lyapunov-function $V(\boldsymbol{x}) = H(\boldsymbol{q}, \boldsymbol{p})$.

According to Theorem 2.1.28, the dissipation inequality for a port-Hamiltonian systems is given by

$$\dot{H}(\boldsymbol{q}, \boldsymbol{p}) \leq \langle \boldsymbol{y}, \boldsymbol{u} \rangle . \tag{2.39}$$

Theorem 2.1.29. *Asymptotic stability of port-Hamiltonian systems [Sch00]*
Suppose a port-Hamiltonian system with a locally stable equilibrium $\boldsymbol{x}^* = [\boldsymbol{q}^*\ 0]^T$ according to Theorem 2.1.28. A negative feedback of the output

$$\boldsymbol{u} = -\boldsymbol{K}_d \cdot \boldsymbol{y}, \quad \boldsymbol{K}_d > 0 \tag{2.40}$$

renders the equilibrium locally asymptotically stable.

By the negative feedback from Theorem 2.1.29, the dissipation inequality from (2.39) results in

$$\dot{H}(\boldsymbol{q}, \boldsymbol{p}) \leq \langle \boldsymbol{y}, -\boldsymbol{K}_d \cdot \boldsymbol{y} \rangle = -\boldsymbol{y}^T \boldsymbol{K}_d \boldsymbol{y} < \boldsymbol{0}. \tag{2.41}$$

This implies strict output-passivity, that is the condition for asymptotic stability according to Theorem 2.1.21.

2.2 Constrained mechanics and nonholonomic systems

This section presents a summary of required equations and some derivations from constrained mechanics in a consistent tensor notation. For more details on constrained mechanics and nonholonomic systems, please refer to the book by Bloch et. al. [Blo16] or the paper by Van der Schaft and Maschke [VM94].

Let $\mathcal{Q}$ be a configuration space with the local coordinates $q^i = \begin{pmatrix} q^1 & \cdots & q^n \end{pmatrix}$, the tangent space $\mathcal{T}$, the co-tangent space $\mathcal{T}^*$, and a smooth Lagrangian function $L : \mathcal{T}\mathcal{Q} \to \mathbb{R}$ denoted by $L(q, p)$, satisfying the regularity condition

$$\det \left[\frac{\partial^2 L}{\partial q_i \, \partial q_j} \right] \neq 0. \tag{2.42}$$

This is for instance satisfied, when the Lagrangian is the kinetic energy with a definite mass matrix. Assume also that κ classical constraints are given in the local coordinates as

$$C_i^a(q) \, \dot{q}^i = 0. \tag{2.43}$$

Assume $\boldsymbol{C}(\boldsymbol{q})$ has the rank κ everywhere . Equation (2.43) defines a κ-dimensional distribution $\boldsymbol{D}$ on $\mathcal{Q}$ given in every point $\boldsymbol{q}_0$ as $\boldsymbol{D}(\boldsymbol{q}_0) = \ker \boldsymbol{C}^T(\boldsymbol{q}_0)$. *Holonomic constraints* (on velocities) are defined to be such that they lead to constraints by integration. In other words, the *integral curves* $q^i(t)$ exist that satisfy (2.43) for all $t \geq 0$. This condition immediately leads to the integrability of $\boldsymbol{D}$. According to the Frobenious theorem, thus, the constraints (2.43) are called holonomic if the distribution $\boldsymbol{D}$ is involutive. In this case, local coordinates $\tilde{q}^j = \Phi^j(q)$ can be found such that

$$
\begin{pmatrix} \dot{\tilde{q}}^{n-\kappa+1} \\ \dot{\tilde{q}}^{n-\kappa+2} \\ \vdots \\ \dot{\tilde{q}}^n \end{pmatrix} = \boldsymbol{0} \Rightarrow \begin{pmatrix} \tilde{q}^{n-\kappa+1} \\ \tilde{q}^{n-\kappa+2} \\ \vdots \\ \tilde{q}^n \end{pmatrix} = \begin{pmatrix} c^{n-\kappa+1} \\ c^{n-\kappa+2} \\ \vdots \\ c^n \end{pmatrix}
\tag{2.44}
$$

for certain constants c^a. The Equations of Motion (EoMs) on the constrained space

$$
\mathcal{D} := \{ (q, p) \in T\mathcal{Q} \,|\, C_i^a(q)\,\lambda_a = 0 \}
\tag{2.45}
$$

are defined by

$$
\frac{d}{dt}\frac{\partial L}{\partial \dot{q}^i} - \frac{\partial L}{\partial q^i} = C_i^a(q)\,\lambda_a \,, \quad C_j^a(q)\,\dot{q}^j = 0
\tag{2.46}
$$

with $C_i^a(q)\,\lambda_a$ the constraint forces and $\boldsymbol{\lambda}(t) \in \mathbb{R}^\kappa$ uniquely determined by requiring that (2.43) is satisfied for all $t \geq 0$.

For deriving the equivalent EoMs, the Hamiltonian $H : T^*\mathcal{Q} \to \mathbb{R}$ is defined by the Legendre transformation

$$
H(q, p) = p_i\,\dot{q}^i - L(q, \dot{q}) \,, \quad p_i = \frac{\partial L}{\partial \dot{q}^i},
\tag{2.47}
$$

and the EoMs (2.46) transform into the constraint Hamiltonian equations on $T^*\mathcal{Q}$

$$
\dot{q}^j = \frac{\partial H(q, p)}{\partial p_j} \,, \quad \dot{p}_i = -\frac{\partial H}{\partial q^i} + C_i^a(q)\,\lambda_a \,, \quad C_i^a(q)\,\frac{\partial H}{\partial p_i} = 0.
\tag{2.48}
$$

Equation (2.48) is obviously a Differential-Algebraic Equation (DAE). The constraint forces $C_i^a(q)\,\lambda_a$ can be calculated by differentiating the constraint condition (2.43) along the trajectories of (2.48):

$$
\frac{d}{dt}\left(C_i^a(q)\,\dot{q}^i \right) = 0
\tag{2.49}
$$

$$
\frac{\partial\, C_i^a \dot{q}^i}{\partial q^j}\,\dot{q}^j + \frac{\partial\, C_i^a \dot{q}^i}{\partial p_j}\,\dot{p}^j = 0
\tag{2.50}
$$

$$
\frac{\partial}{\partial q^j}\left(C_i^a \frac{\partial H}{\partial p_i} \right)\frac{\partial H}{\partial p_j} + \frac{\partial}{\partial p_j}\left(C_i^a \frac{\partial H}{\partial p_i} \right)\left[-\frac{\partial H}{\partial q^j} + C_j^a(q)\,\lambda_a \right] = 0
\tag{2.51}
$$

$$
\frac{\partial}{\partial q^j}\left(C_i^a \frac{\partial H}{\partial p_i} \right)\frac{\partial H}{\partial p_j} + C_i^a \frac{\partial^2 H}{\partial p_i \partial p_j}\left[-\frac{\partial H}{\partial q^j} + C_j^a(q)\,\lambda_a \right] = 0
\tag{2.52}
$$

This algebraic equation can be solved at every t to determine λ_a as a function of q and p. Substituting into (2.48) yields the EoMs on the *constraint state space* $\mathcal{M}$ defined by

$$\mathcal{M} := \left\{ (q,p) \in \mathcal{T}^* \mathcal{Q} \,|\, C_i^a(q)\, \frac{\partial H}{\partial p_i} = 0 \right\} \tag{2.53}$$

Using the structure matrix $\boldsymbol{J}$

$$\boldsymbol{J} := \begin{pmatrix} \boldsymbol{0}_n & \boldsymbol{I}_n \\ -\boldsymbol{I}_n & \boldsymbol{0}_n \end{pmatrix} \tag{2.54}$$

the equations (2.48) can also be rewritten as

$$\begin{pmatrix} \dot{\boldsymbol{q}} \\ \dot{\boldsymbol{p}} \end{pmatrix} = \boldsymbol{J} \begin{pmatrix} \dfrac{\partial H(\boldsymbol{q},\boldsymbol{p})}{\partial \boldsymbol{q}} \\ \dfrac{\partial H(\boldsymbol{q},\boldsymbol{p})}{\partial \boldsymbol{p}} \end{pmatrix} + \begin{pmatrix} \boldsymbol{0} \\ \boldsymbol{C}(\boldsymbol{q}) \end{pmatrix} \lambda \,, \quad \boldsymbol{C}^T(\boldsymbol{q})\, \frac{\partial H(\boldsymbol{q},\boldsymbol{p})}{\partial \boldsymbol{p}} = \boldsymbol{0}. \tag{2.55}$$

As stated in [Blo16], the EoMs on the constrained spaces can also be determined in a more intrinsic way considering the constraints already in the derivation. This leads to a set of Ordinary Differential Equations (ODEs), contrary to the DAE above, that can be used e.g. for control synthesis and simulations. The result is basically the EoMs as from (2.46) or (2.48) where the constraint forces are eliminated.

Consider as before a mechanical system on a configuration space $\mathcal{Q}$ with n generalised coordinates q^i subject to κ nonholonomic constraints given by (2.43). Assume choosing a set of local coordinates r and s with

$$q = \begin{bmatrix} \boldsymbol{r} \\ \boldsymbol{s} \end{bmatrix} \text{ or } q^i = [r^\alpha, \, s^a] \tag{2.56}$$

such that the constraints are described using an Ehresmann-Connection A, that is

$$C_i^a\, \dot{q}^i = \dot{s}^a + A_\alpha^a\, \dot{r}^\alpha = 0 \,. \tag{2.57}$$

In the literature, r^α are often referred to as the coordinates of the *horizontal space* or are called the *base* coordinates and s^a are referred to as the coordinates of the *vertical space* or are called the *fiber* coordinates.

The Lagrange-d'Alembert equations of motion are given by the variation

$$-\delta L = \left(\frac{d}{dt}\, \frac{\partial L}{\partial \dot{q}^i} - \frac{\partial L}{\partial q^i} \right) \delta q^i = 0 \tag{2.58}$$

where the variations $\delta q^i(t)$ are chosen such that the integral curves $q^i(t)$ satisfy the constraints. The variation of (2.57) delivers

$$\delta s^a + A_\alpha^a\, \delta r^\alpha = \omega^a = 0 \quad \Rightarrow \quad \delta s^a = -A_\alpha^a\, \delta r^\alpha. \tag{2.59}$$

The variation equation (2.58) is then

$$\frac{d}{dt}\frac{\partial L}{\partial \dot{q}^i}\,\delta q^i - \frac{\partial L}{\partial q^i}\,\delta q^i = 0$$

$$\Rightarrow \frac{d}{dt}\frac{\partial L}{\partial \dot{r}^\alpha}\,\delta r^\alpha + \frac{d}{dt}\frac{\partial L}{\partial \dot{s}^a}\,\delta s^a - \frac{\partial L}{\partial r^\alpha}\,\delta r^\alpha - \frac{\partial L}{\partial s^a}\,\delta s^a = 0$$

$$\Rightarrow \left(\frac{d}{dt}\frac{\partial L}{\partial \dot{r}^\alpha} - \frac{\partial L}{\partial r^\alpha}\right)\delta r^\alpha = -\left(\frac{d}{dt}\frac{\partial L}{\partial \dot{s}^a} - \frac{\partial L}{\partial s^a}\right)\delta s^a. \tag{2.60}$$

Substituting (2.59) yields

$$\left(\frac{d}{dt}\frac{\partial L}{\partial \dot{r}^\alpha} - \frac{\partial L}{\partial r^\alpha}\right)\delta r^\alpha = \left(\frac{d}{dt}\frac{\partial L}{\partial \dot{s}^a} - \frac{\partial L}{\partial s^a}\right)A^a_\alpha\,\delta r^\alpha \tag{2.61}$$

$$\Rightarrow \left(\frac{d}{dt}\frac{\partial L}{\partial \dot{r}^\alpha} - \frac{\partial L}{\partial r^\alpha}\right) = \left(\frac{d}{dt}\frac{\partial L}{\partial \dot{s}^a} - \frac{\partial L}{\partial s^a}\right)A^a_\alpha \tag{2.62}$$

$$\Rightarrow \frac{d}{dt}\frac{\partial L}{\partial \dot{r}^\alpha} - \frac{\partial L}{\partial r^\alpha} = A^a_\alpha\left(\frac{d}{dt}\frac{\partial L}{\partial \dot{s}^a} - \frac{\partial L}{\partial s^a}\right) \tag{2.63}$$

Define now the *constraint Lagrangian* $L_\mathcal{D}$ as

$$L_\mathcal{D}(r^\alpha, s^a, \dot{r}^\alpha) := L(r^\alpha, s^a, \dot{r}^\alpha, \underbrace{-A^a_\alpha\,\dot{r}^\alpha}_{\dot{s}^a}) \tag{2.64}$$

and also calculate the following derivatives

$$\frac{\partial L_\mathcal{D}}{\partial \dot{r}^\alpha} = \frac{\partial L}{\partial \dot{r}^\alpha} - A^b_\alpha \frac{\partial L}{\partial \dot{s}^b} \qquad\Rightarrow\qquad \frac{\partial L}{\partial \dot{r}^\alpha} = \frac{\partial L_\mathcal{D}}{\partial \dot{r}^\alpha} + A^b_\alpha \frac{\partial L}{\partial \dot{s}^b} \tag{2.65}$$

$$\frac{\partial L_\mathcal{D}}{\partial r^\alpha} = \frac{\partial L}{\partial r^\alpha} - \frac{\partial L}{\partial \dot{s}^b}\left(\frac{\partial A^b_\beta}{\partial r^\alpha}\dot{r}^\beta\right) \qquad\Rightarrow\qquad \frac{\partial L}{\partial r^\alpha} = \frac{\partial L_\mathcal{D}}{\partial r^\alpha} + \frac{\partial L}{\partial \dot{s}^b}\frac{\partial A^b_\beta}{\partial r^\alpha}\dot{r}^\beta \tag{2.66}$$

$$\frac{\partial L_\mathcal{D}}{\partial s^a} = \frac{\partial L}{\partial s^a} - \frac{\partial L}{\partial \dot{s}^b}\left(\frac{\partial A^b_\beta}{\partial s^a}\dot{r}^\beta\right) \qquad\Rightarrow\qquad \frac{\partial L}{\partial s^a} = \frac{\partial L_\mathcal{D}}{\partial s^a} + \frac{\partial L}{\partial \dot{s}^b}\frac{\partial A^b_\beta}{\partial s^a}\dot{r}^\beta. \tag{2.67}$$

Substituting (2.65), (2.66) and (2.67) into (2.63) means to consider the constraints *after* taking the variation. This leads to

$$\frac{d}{dt}\left(\frac{\partial L_{\mathcal{D}}}{\partial \dot{r}^\alpha} + A^b_\alpha \frac{\partial L}{\partial \dot{s}^b}\right) - \frac{\partial L_{\mathcal{D}}}{\partial r^\alpha} - \frac{\partial L}{\partial \dot{s}^b}\frac{\partial A^b_\beta}{\partial r^\alpha}\dot{r}^\beta = A^a_\alpha\left(\frac{d}{dt}\frac{\partial L}{\partial \dot{s}^a} - \frac{\partial L_{\mathcal{D}}}{\partial s^a} - \frac{\partial L}{\partial \dot{s}^b}\frac{\partial A^b_\beta}{\partial s^a}\dot{r}^\beta\right) \qquad (2.68)$$

$$= A^a_\alpha\frac{d}{dt}\frac{\partial L}{\partial \dot{s}^a} - A^a_\alpha\frac{\partial L_{\mathcal{D}}}{\partial s^a} - A^a_\alpha\frac{\partial L}{\partial \dot{s}^b}\frac{\partial A^b_\beta}{\partial s^a}\dot{r}^\beta. \qquad (2.69)$$

Therefore

$$\frac{d}{dt}\frac{\partial L_{\mathcal{D}}}{\partial \dot{r}^\alpha} - \frac{\partial L_{\mathcal{D}}}{\partial r^\alpha} + A^a_\alpha\frac{\partial L_{\mathcal{D}}}{\partial s^a}$$

$$= -\frac{d}{dt}\left(A^b_\alpha\frac{\partial L}{\partial \dot{s}^b}\right) + A^a_\alpha\frac{d}{dt}\frac{\partial L}{\partial \dot{s}^a} - A^a_\alpha\frac{\partial L}{\partial \dot{s}^b}\frac{\partial A^b_\beta}{\partial s^a}\dot{r}^\beta + \frac{\partial L}{\partial \dot{s}^b}\frac{\partial A^b_\beta}{\partial r^\alpha}\dot{r}^\beta \qquad (2.70)$$

$$= -\frac{d}{dt}A^b_\alpha\frac{\partial L}{\partial \dot{s}^b}\underbrace{- A^b_\alpha\frac{d}{dt}\frac{\partial L}{\partial \dot{s}^b} + A^a_\alpha\frac{d}{dt}\frac{\partial L}{\partial \dot{s}^a}}_{=0} - A^a_\alpha\frac{\partial L}{\partial \dot{s}^b}\frac{\partial A^b_\beta}{\partial s^a}\dot{r}^\beta + \frac{\partial L}{\partial \dot{s}^b}\frac{\partial A^b_\beta}{\partial r^\alpha}\dot{r}^\beta \qquad (2.71)$$

$$= -\frac{d}{dt}A^b_\alpha\frac{\partial L}{\partial \dot{s}^b} - A^a_\alpha\frac{\partial L}{\partial \dot{s}^b}\frac{\partial A^b_\beta}{\partial s^a}\dot{r}^\beta + \frac{\partial L}{\partial \dot{s}^b}\frac{\partial A^b_\beta}{\partial r^\alpha}\dot{r}^\beta \qquad (2.72)$$

$$= -\frac{d}{dt}A^b_\alpha\frac{\partial L}{\partial \dot{s}^b} + \frac{\partial L}{\partial \dot{s}^b}\left(-A^a_\alpha\frac{\partial A^b_\beta}{\partial s^a} + \frac{\partial A^b_\beta}{\partial r^\alpha}\right)\dot{r}^\beta. \qquad (2.73)$$

On the other hand, the time derivative of the tensor A is

$$\frac{d}{dt}A^b_\alpha = \frac{\partial A^b_\alpha}{\partial s^a}\dot{s}^a + \frac{\partial A^b_\alpha}{\partial r^\beta}\dot{r}^\beta = -\frac{\partial A^b_\alpha}{\partial s^a}A^a_\beta\dot{r}^\beta + \frac{\partial A^b_\alpha}{\partial r^\beta}\dot{r}^\beta \qquad (2.74)$$

$$\Rightarrow \left(\frac{\partial A^b_\alpha}{\partial r^\beta} - \frac{\partial A^b_\alpha}{\partial s^a}A^a_\beta\right)\dot{r}^\beta = -\left(A^a_\beta\frac{\partial A^b_\alpha}{\partial s^a} - \frac{\partial A^b_\alpha}{\partial r^\beta}\right)\dot{r}^\beta \qquad (2.75)$$

that turns (2.73) into

$$\frac{d}{dt}\frac{\partial L_{\mathcal{D}}}{\partial \dot{r}^\alpha} - \frac{\partial L_{\mathcal{D}}}{\partial r^\alpha} + A^a_\alpha\frac{\partial L_{\mathcal{D}}}{\partial s^a} = -\frac{d}{dt}A^b_\alpha\frac{\partial L}{\partial \dot{s}^b} + \frac{\partial L}{\partial \dot{s}^b}\left(-A^a_\alpha\frac{\partial A^b_\beta}{\partial s^a} + \frac{\partial A^b_\beta}{\partial r^\alpha}\right)\dot{r}^\beta \qquad (2.76)$$

$$= \frac{\partial L}{\partial \dot{s}^b}\left(A^a_\beta\frac{\partial A^b_\alpha}{\partial s^a} - \frac{\partial A^b_\alpha}{\partial r^\beta} + \frac{\partial A^b_\beta}{\partial r^\alpha} - A^a_\alpha\frac{\partial A^b_\beta}{\partial s^a}\right)\dot{r}^\beta \qquad (2.77)$$

Finally, defining the tensor B - which is called the *curvature* of A in the literature - the μth row of the EoMs on the constrained space is given by

$$\frac{d}{dt}\frac{\partial L_\mathcal{D}}{\partial \dot{r}^\mu} - \frac{\partial L_\mathcal{D}}{\partial r^\mu} + A_\mu^a \frac{\partial L_\mathcal{D}}{\partial s^a} = -\frac{\partial L}{\partial \dot{s}^b} B_{\mu\alpha}^b \dot{r}^\alpha + \tau_\mu ,\tag{2.78}$$

$$B_{\mu\alpha}^b := -\left(\frac{\partial A_\mu^b}{\partial s^a}A_\alpha^a - \frac{\partial A_\mu^b}{\partial r^\alpha} + \frac{\partial A_\alpha^b}{\partial r^\mu} - \frac{\partial A_\alpha^b}{\partial s^a}A_\mu^a\right).\tag{2.79}$$

Note that the external generalised forces τ_μ are added to (2.78) to complete the EoM.

For the derivation of the equivalent equations in the Hamiltonian framework, a coordinate transformation is applied. Since the rank of $\boldsymbol{C}$ is κ, there exists locally an annihilator which is a smooth $n \times (n-\kappa)$ tensor S of rank $n-\kappa$ such that

$$C_i^a(q)\,S_\alpha^i(q) = 0 \in \mathbb{R}^{\kappa\times(n-\kappa)}\tag{2.80}$$

The new impulse coordinates $\rho_i = \begin{bmatrix}\rho_\alpha \\ \rho_a\end{bmatrix}$ are defined as

$$\rho_\alpha := S_\alpha^i(q)\,p_i ,\quad \alpha \in \{1,\cdots,n-\kappa\}$$
$$\rho_a := C_a^i(q)\,p_i ;\quad a \in \{n-\kappa+1,\cdots,n\}.\tag{2.81}$$

In the literature, ρ_α and ρ_a are often referenced to as *base-impulse-coordinates* and *fiber-impulse-coordinates* correspondingly. Note that $((q,p) \to (q,\rho))$ is a coordinate transformation because of (2.80) since the entire configuration space is covered by C and S. Consider the Ehresmann connection A from (2.59) that leads to

$$\dot{s}^a + A_\alpha^a \dot{r}^\alpha = 0 \;\Rightarrow\; \begin{bmatrix}A_\alpha^a & \delta_b^a\end{bmatrix}\begin{bmatrix}\dot{r}^\alpha \\ \dot{s}^b\end{bmatrix} = \mathbf{0} \;\Rightarrow\; C_j^a = \begin{bmatrix}A_\alpha^a & \delta_b^a\end{bmatrix}\tag{2.82}$$

$$C_i^a(q)\,S_\alpha^i(q) = 0 \;\Rightarrow\; S_\alpha^i = \begin{bmatrix}\delta_\alpha^\beta & -A_\alpha^a\end{bmatrix}\tag{2.83}$$

with δ being the Kronecker-Delta representing a unity matrix in the tensor notation. Taking $p_i = \begin{bmatrix}p_\alpha & p_a\end{bmatrix}$ corresponding to the split of q from (2.56), the transformed base and fiber momenta are given by

$$\rho_\alpha = p_\alpha - A_\alpha^a p_a\tag{2.84}$$
$$\rho_a = A_a^\alpha p_\alpha + p_a = p_a + A_a^\alpha p_\alpha\tag{2.85}$$

The Hamiltonian equations for the (fully actuated) system can be given in the well-known matrix representation as ([VM94])

$$\begin{pmatrix}\dot{q}^i \\ \dot{\rho}_\alpha\end{pmatrix} = \begin{pmatrix}0 & S_\alpha^i(q) \\ -S_\alpha^i(q) & R_{\alpha\beta}(q,p)\end{pmatrix}\begin{pmatrix}\dfrac{\partial H_\mathcal{M}(q,\rho)}{\partial q^i} \\ \dfrac{\partial H_\mathcal{M}(q,\rho)}{\partial \rho_\beta}\end{pmatrix} + \begin{pmatrix}0 \\ \tau_\alpha\end{pmatrix}\tag{2.86}$$

with

$$R_{\alpha\beta} = -B_{\alpha\beta}^c \, p_c \, , \quad S_\alpha^i = \begin{pmatrix} \delta_\alpha^\beta \\ -A_\alpha^b \end{pmatrix} \, , \quad \dot{q}^i = \begin{pmatrix} \dot{r}^\alpha \\ \dot{s}^a \end{pmatrix}. \tag{2.87}$$

The equations above represent a so-called *pseudo-Hamiltonian* system as stated in [VM94] since the constraints given by the Ehresmann-Connection A are assumed to be nonholonomic. However, these equations can be still used for the purpose of controller design as shown in [FSS04]. As proven in [KM97], the equations of motion (2.78) and (2.86) are equivalent. For numeric implementation of the EoMs it is worth to note that p_c in (2.87) is basically the part of the original impulse coordinates corresponding to the fiber-coordinates. Given only the transformed impulse coordinates, p_c is obtained by

$$p_c = \begin{bmatrix} 0_\alpha^i & \delta_a^i \end{bmatrix} M_{ij} \dot{q}^j = \begin{bmatrix} 0_\alpha^i & \delta_a^i \end{bmatrix} M_{ij} \, S_\beta^j \, M^{D,\beta\gamma} \, \rho_\gamma. \tag{2.88}$$

3 Physics of two-wheeled vehicles

Physical phenomena involved in the behaviour of Two-Wheeled Vehicles (TWVs) are quite interesting. Although most people use bicycles regularly, some of those phenomena are either not paid attention to or even misconceived. Yet, those are very important with regard to mathematical modelling and systems theoretical analysis of a TWV as a dynamical system. In this chapter, the most important physical characteristics of a TWV are explained briefly.

In Section 3.1 the steering behaviour of bicycles is addressed, and, elaborated how bicycle are steered by a combination of the leaning angel and the steering angle. The so-called counter-steering behaviour of bicycles and the steerability of TWVs depending on the leaning angle are further explained. In Section 3.2, the self-stability of bicycles and its origins are explained. In Section 3.3, the so-called linear benchmark model of bicycles is discussed, and, the fact, how self-stability of bicycles is explained by eigenvalue analysis of a linear system.

3.1 Steering behaviour

Contrary to four-wheeled vehicles, steering of a TWV is not done by only changing the steering angle. Generally speaking, bicycles are steered by a combination of steering and leaning. For instance in a left-curve, the rider steers to the left and leans the vehicle to the left, too. The interesting fact is that, in contrast to mobile robots with Ackermann-steering for instance, there is not a unique steering angle corresponding to a specific curve radius. Depending on the vehicle velocity, and thus the centripetal force acting on the vehicle, there are many combinations of steering and leaning angle to keep the vehicle in a curve with a specific radius.

Non-minimum-phase behaviour
One of the characteristics of bicycles that is mostly unnoticed is the so-called *counter-steering*. It is much better known among motorcycle riders, since at higher velocities and with a larger vehicle mass, the counter-steering has to be considered for steering the vehicle. In particular, for entering a left curve for instance, the vehicle has to lean and steer to the left. However, starting with a straight ride, steering to the left causes the vehicle to lean to the right. Then, to avoid falling over, the rider will have to steer to the right. Therefore to enter a left-curve, the rider has to steer to the opposite direction, namely right, first. This fact is very well explained in a video from MinutePhysics, from which a screenshot is show in

Figure 3.1 In a systems theoretical context, taking the steering angel as input of the system and the vehicle orientation or the leaning angle as the output, this is a non-minimum-phase behaviour.

Steerability

The yaw-rate of TWVs does not directly depend on the steering angel. As explained before, the leaning angle is also involved in the change of the vehicle orientation. In particular, the change of the orientation depends on the so-called *real-steering angel*, that is the projection of the steering axis on the ground. As will be explained mathematically more precisely later, the larger the leaning angle, the more sensitive is the yaw change to the change of the steering axis. Therefore, the vehicle is also less steerable with a larger steering angle. This fact is used by racing motorcycle pilots leaning their body into the curve to reduce the required steering angle and, therefore, the steerability of the vehicle at high velocities in a curve as illustrated in the picture in Figure 3.3

3.2 Self-stability

Maybe the most fascinating phenomenon observed in TWVs is the so-called *self-stability*. That is, in a particular range of the vehicle velocity and depending on the physical entities such as mass and geometry, a rider-less TWV stays upright and keeps moving even in presence of external disturbances. This phenomenon has been noticed from the beginning of the historical development of bicycles as early as 1886 according to [MP11]. As described in [MP11], there are many effects that are believed to be involved in the self-stability of bicycles. However, which one is necessary and which one is sufficient as well as the right combination of the effects is still unknown. In general, self-stability is a result of the rider-less bicycle steering to the *proper direction*, once the vehicle is leaned to a side. Three main geometric entities are commonly accepted to be involved in the self-stability of TWVs.

- **Trail:** The positive trail, that is the distance between the front wheel contact point and the intersection of the ground plane with the extension of the steering axis, is the mostly mentioned geometric entity in the literature. Due to the trail, the ground contact forces cause steering to one side once the vehicle is leaned to the same side.

- **Mass distribution:** The mass distribution of the steering mechanism also causes proper steering due to gravity and centripetal forces, especially once the centre of mass of the steering mechanism is slightly in front of the steering axis.

- **Gyroscopic force:** The gyroscopic force due to the rotation of the front wheel also leads to a steering caused by the leaning of the vehicle.

This fact is explained well in the video from MinutePhysics, from which a screenshot is shown in Figure 3.2. Later in this thesis, these effects are taken into account in the mathematical modelling of the vehicle and the self-stability is investigated by simulations.

Counter-Steering (https://youtu.be/l1Rkf1fnNDM)

1. Steer to the right   2. Lean to the left   3. Steer to the left

Figure 3.1: Screenshot from the MinutePhysics video: How bicycles turn to the left

Self-Stability (https://youtu.be/oZAc5t21kvo)

1. Positive trail   2. Mass distribution   3. Gyroscopic force

Figure 3.2: Screenshot from the MinutePhysics video: Why are bicycles self-stable?

Figure 3.3: Riders leaning into the curve to increase steerability (www.cycleworld.com)

3.3 The linear benchmark model of bicycles

The so-called *linear benchmark model* (also Whipple-model) provides a systematic analysis
of the bicycle motion through establishing four distinct eigenmodes, corresponding to the
eigenvalues of a linearised bicycle model. Note that the benchmark model is a Linear
Parameter-Varying (LPV) system with four states, namely the leaning angle, the steering
angle and the corresponding angular rates. In this sense, the term stability refers to the
steering angle and (mainly) the leaning angel of the bicycle. That is, loosely speaking, the
bicycle in the upright pose does not fall over due to a disturbance.

The picture illustrated in Figure 3.4 is from [MPRS07] that shows the change of the eigenvalues with respect to the vehicle velocity v on the horizontal axis. The vertical axis
corresponds to the eigenvalues, where the real parts and the imaginary parts are pictured
separately. Three eigenmodes are are established, that are the weave mode, the capsize
mode and the castering mode.

The eigenvalues of both the castering and the capsize mode are purely real for every velocity.
At very low velocities $v < v_d$, these eigenvalues are both negative. In this velocity range,
a bicycle behaves like an inverted pendulum, since the two eigenvalues corresponding to
the weave mode are real and positive. At higher velocities $v > v_d$, the eigenvalues of the
weave mode are a conjugate pair. The weave mode is characterised by the oscillations in
both the lean and the steer rates with steering rate having a phase lag relative to leaning
[MPRS07, p.15]. The frequency of the oscillations increases with increasing speeds indicated
by increasing the absolute values of the imaginary part. With the increasing velocity, the
real part of the eigenvalues of the weave mode is reduced, until they become negative for
$v > v_w$. This is the lower bound of the self-stability range since the other two eigenvalues are
still negative. Here, the magnitude of the oscillations caused by the weave mode converges to
zero, as if there were a damping involved. If the velocity is further increased, the eigenvalue
of the capsize mode approaches zero, and, for $v > v_c$ it becomes positive. Therefore the
self-stability range is bounded by $v_d < v < v_c$. For velocities $v > v_c$, the bicycle is no more
self-stable. In fact, with increasing velocity, the bicycle appears to behave similar the to
case of $v < v_d$, since the frequency of the weave mode oscillations is too high.

The nonlinear model for TWVs introduced in Chapter 4 is compared to the benchmark
model for verification, both in simulations and by analysing eigenvalues of the linearised
model. Note that this model, also called the Whipple model, is referred to as the *benchmark
model* in the literature and this phrase is taken over in this work. This, however, does not
mean that the exact behaviour of this model is taken as an exact numerical benchmark for
comparing to the introduced model. For instance, the eigenvalues of the linear system in
Section 4.5.2 are not the same as shown in Figure 3.4.

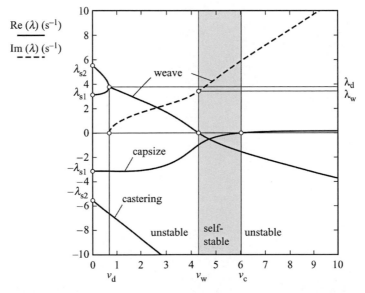

Figure 3.4: Eigenvalues λ of the benchmark bicycle model from [MPRS07]

4 Lagrangian and Hamiltonian modelling of two-wheeled vehicles

In this chapter, the formalisms explained in Section 2.2 are used to derive a generalised workflow for obtaining the EoMs for constrained mechanical systems using metric tensors. The resulted scheme is well-structured and allows automation using software tools as well as straight forward extension of the model of a mechanical system by a new rigid body. In addition to the tensors of the EoMs for a self-stable TWV in the Lagrangian form, the required tensors for the pseudo-Hamiltonian equations (2.86) are obtained, that are later used for controller synthesis in Chapter 6.

Section 4.1 presents a review on the existing bicycle models from the literature, with which the introduced model in Section 4.4 is classed.

In Section 4.2, known formalisms are reformulated to obtain an automatable sequence for obtaining Lagrangian and Hamiltonian EoMs for constrained mechanical systems. Thereby, the definition and the properties of metric tensors are used. The proposed steps can be applied for multi-body mechanical system with holonomic or nonholonomic constraints.

The main result of this chapter, the proposed TWV model in Section 4.4, is too complex and the corresponding tensors cannot be written out in full. Therefore, the procedure is demonstrated in Section 4.3, where a simple mechanical system is considered to elaborate the steps, and the resulting tensors in a clear and well-readable form.

In Section 4.4, some geometric considerations and change of coordinate systems make sure, that the physical phenomena leading to a realistic behaviour of TWVs - as explained in Chapter 3 - are taken into account while deriving the vehicle model. It is assumed that the vehicle is a 2-body system, consisting of a rear frame and a steering assembly. However, to complete the main physical phenomena known to affect bicycles' self-stability, the gyroscopic force on the steering axis due to the rotation of the front wheel is considered, too. As mentioned before, due to the complexity and the size of the terms involved in the TWV model, only the necessary calculations are written out and the resulting tensors are omitted, which can be found in a referenced online repository.

The validation of the resulting model is investigated in two different ways in this chapter. In Section 4.5.1, a set of simulation scenarios is devised, each for verification of a specific physical phenomenon, which are explained in Chapter 3. Thereby, it is examined whether a physically expected behaviour pattern is consistent with the simulation results. Specifically, the simulation results verify the consistency of the vehicle model with regard to the balancability at zero forward velocity, the self-alignment toque on the steering mechanism and the self-stability in a certain velocity range. The counter steering behaviour is also verified as well as the fact, that a moving bicycle is steerable both through the steering torque and moving of the riders body - that is the hands-free riding of a bicycle.

In Section 4.5.2, the proposed model is validated by a comparison to the so-called linear benchmark model. As explained in Section 3.3, the benchmark model is a LPV system with the constant forward velocity v as the varying parameter, which is resulted by linearising a rather sophisticated four-body bicycle model. The linear benchmark model is well-known to represent the self-stability behaviour of bicycles and is characterised by the eigenvalues of the system matrix. Therefore, the proposed model is linearised for a constant forward velocity v and the eigenvalues of resulting linear system are analysed. A comparison shows that the eigenvalues follow the same pattern as those for the benchmark mode. Also, simulation results from Section 4.5 verify that the corresponding eigenmodes are consistent with those for the benchmark model.

For some time-critical applications such as online trajectory planning, the proposed vehicle model may be too complex and lead to too large computation times. Therefore and using the structure of the resulting equations from Section 4.2, in Section 4.6, a model simplification is proposed by approximating the tensors involved in the model. It is shown that approximating the tensors using the Taylor series up to the linear term results in a simplified model with satisfyingly similar behaviour to the original model. Another simplified model, which shows a closer behaviour to the original model while preserving the port-Hamiltonian structure of the system, is obtained by a quadratic approximation of the metric tensor.

An important advantage of the scheme introduced in Section 4.2 is - beside being automatable using symbolic computer tools - the extendability. This means, that a new rigid body can be added to a mechanical system without a significant change in the script of the symbolic computer tool. Further in Section 4.6, this fact is demonstrated by adding an actively leaning rider to the bicycle model as an example for an extension.

At the end of this chapter, the effects of the idealisation, especially with respect to friction and tier's lateral slip are discussed briefly.

4.1 Existing models for two-wheeled vehicles

With respect to the model based control, often a trade-off is necessary between model accuracy and usability. As explained in Chapter 3, bicycles are known to be self-stable at a particular velocity that is, e.g., discussed in [AKL05]. The most important existing bicycle model, the so-called linear benchmark model, is described in [MPRS07], which is a LPV system with the forward velocity v as varying parameter, and is based on the work of Whipple [Whi99]. The authors of [MPRS07] state that the self-stability of the uncontrolled bicycle is due to several geometric, inertial and gyroscopic features[1]. That paper laid the foundations for a systematic analysis of the bicycle dynamics and, most interestingly, the

---

[1]This is also explained very well in the video from MinutePhysics: `youtu.be/oZAc5t2lkvo`

self-stability. The model was experimentally validated, for instance by [KSM08].

Since that model is only valid for constant velocities, [LS08] added acceleration by extending the linear model. The self-stability of the vehicle depending on geometric design parameters was investigated by [TWB10]. There is also insightful work on motorcycle dynamics, mostly for purposes of vehicle design, e.g. by [Cos06]. It is known that the bicycle is self-stable at a certain range of the forward velocity. This leads to the fact that a bicycle is controllable by two different actuations, the steering and the leaning of the rider. This was analysed by [CP10] and further investigated by [SMK12]. An overall review is given by [SM13], which is currently the most comprising summary on bicycle models known to the author. The authors of [SM13] states that rides on curves with small radii cannot be described by the linearised model. A review on bicycle models that focus more on the rider is given by [KS13].

The linear benchmark model has already been used to design controllers, for instance by [KMT09]. In [EHP15], an inverted pendulum is added to the model to represent a passive rider and designed a controller. The authors of [KYKY11] consider a Control Momentum Gyro (CMG) to apply direct torque on the leaning angle and verify their results experimentally.

Meijaard et. al. [MPRS07] divides existing models into three groups. First group contains models derived by qualitative dynamic discussions. The second group of models is obtained using analysis to study the dynamics. The models from both categories are mostly too reduced to cover the self-stability ob bicycles. Some papers categorised in the second group include a control law for the steering to achieve self-stability, for instance [AKL05] and [LS08] to name two of the newest. The third category contains rigid body dynamic models which are detailed enough with regard to the geometry and the mass distribution, for instance [BMCP07].

Among the papers in the second group, the nonlinear model from [GM95] is mentioned whose modelling approach, however, may be categorised in a dedicated new category. In fact, [GM95] uses a special approach of modelling nonholonomic systems using the constrained Lagrangian. As a result, the so-called *Naive* bicycle model is introduced, that is a nonlinear system used to design a nonlinear controller. The *Naive* model, however does not cover the geometry or the mass distribution leading to the self-stability. It was followed by another group of publications, for instance [KM97] which adds the Hamiltonian approach to derive an equivalent model. Zhang et. al. [ZLYS11] adds a term to the potential energy to make the bicycle balanceable at zero forward velocity $v = 0$. The authors of [WYLZ17] added a CMG for active balancing.

The majority of the existing models focus on bicycle design and, analysis of controlled and uncontrolled vehicles. Existing linear models describe the vehicle dynamics for a constant forward velocity. Existing nonlinear models, in contrast, are too complex to be used for systematic control synthesis. Thus, designing a controller, for instance, requires a dedicated synthesis as in [GM95]. The model in this work however, is derived concerning autonomous driving two-wheeled vehicles and corresponding model-based approaches.

The main objective is to develop a model that is sophisticated enough to describe the dynamics of the vehicle. At the same time, it needs to be well-structured and simple enough to be used for tasks such as model-based controller design and optimal trajectory planning. The approach used in this thesis is based on an extension of [GM95] and [KM97]. In contrast to [ZLYS11] however, the shift of the centre of mass of the bicycle due to steering is considered in the geometry, and not as an extra term in the potential energy. This leads to the model being physically general and consistent with the model from [ZLYS11] regarding the balance at $v = 0$. Also, the model is consistent with the linear models discussed above regarding the self-stability and eigenmodes of the motion. In this sense, this is an attempt to bring together the approaches from the two groups of literature mentioned above.

The main requirement of the model is to be generally valid, i.e., it has to represent the vehicle dynamics for every set of parameters, especially for every value of the forward velocity v. The model needs to include the nonholonomic constraints as well as the phenomena leading to self-stability. Furthermore, the model shall be systematically extendible for adding a rider as well as not considered effects such as friction and tire models. Finally, the model shall have a proper structure to allow systematic approaches for controller and observer design.

To fulfil these requirements, the EoMs are derived in the Lagrangian and the Hamiltonian framework. Port-Hamiltonian systems have been approved to be suitable for different purposes and are becoming more important in the branch of nonlinear control, as will be explained in Chapter 6. In general, once a system is defined as a port-Hamiltonian system, systematic synthesis approaches can be applied directly [vdSJ14]. The controller design based on the model is explained in Chapter 6.

4.2 General model derivation for nonholonomic vehicles using the metric tensor

For sake of lucidity, first assume that the Lagrangian contains only the kinetic energy $L = T = \frac{1}{2} M_{ij} \dot{q}^i \dot{q}^j$, where M is the metric tensor on $\mathcal{Q}$. The potential energy will be considered further below. The reduced Lagrangian $L_\mathcal{D}$ has the same properties and can therefore be given in the same form as $L_\mathcal{D} = \frac{1}{2} M_{\mathcal{D},\alpha\beta} \dot{r}^\alpha \dot{r}^\beta$. The tensor $M_\mathcal{D}$ is the reduced metric tensor corresponding to the constrained sub-manifold $\mathcal{D}$ on which the constraints are satisfied. Defining S as in (2.83) and substituting yields

$$\dot{q}^i = [\dot{r}^\alpha , \dot{s}^a] := S_\alpha^i \, \dot{r}^\alpha \, , \;\; L_\mathcal{D} = \frac{1}{2} M_{ij} \, S_\alpha^i \, \dot{r}^\alpha \, S_\beta^j \, \dot{r}^\beta = \frac{1}{2} M_{ij} \, S_\alpha^i \, S_\beta^j \, \dot{r}^\alpha \, \dot{r}^\beta \qquad (4.1)$$

that results in the definition of the reduced metric tensor (mass matrix) as

$$M_{\mathcal{D},\alpha\beta} := M_{ij} \, S_\alpha^i \, S_\beta^j \, . \qquad (4.2)$$

As stated in [KM97], in many mechanical systems such as the bicycle, $A = A(r)$ holds. Also, $M_{\mathcal{D}} = M_{\mathcal{D}}(r)$ by definition. The specific terms of (2.78) are determined as

$$\frac{d}{dt}\frac{\partial L_{\mathcal{D}}}{\partial \dot{r}^\mu} = \frac{d}{dt}[M_{\mathcal{D},\alpha\mu}\,\dot{r}^\alpha] \tag{4.3}$$

$$\frac{\partial L_{\mathcal{D}}}{\partial r^\mu} = \frac{1}{2}\Gamma_{\mathcal{D},\alpha\beta\mu}\,\dot{r}^\alpha\,\dot{r}^\beta\,,\quad \Gamma_{\mathcal{D},\alpha\beta\mu} := \frac{\partial M_{\mathcal{D},\alpha\beta}}{\partial r^\mu} \tag{4.4}$$

$$\frac{\partial L}{\partial \dot{s}^b} = [0_b^\gamma\,,\,\delta_b^a]\,\frac{\partial L}{\partial \dot{q}^i} = [0_b^\gamma\,,\,\delta_b^a]\,M_{ij}\,\dot{q}^j = \tilde{I}_b^i\,M_{ij}\,S_\alpha^j\,\dot{r}^\alpha = M_{\mathcal{S},b\alpha}\,\dot{r}^\alpha \tag{4.5}$$

$$\tilde{I}_b^i := [0_b^\gamma\,,\,\delta_b^a],\quad M_{\mathcal{S},b\alpha} := \tilde{I}_b^i\,M_{ij}\,S_\alpha^j. \tag{4.6}$$

Note that the term $M_{\mathcal{S}}$ is only denoted similarly to a metric tensor because of its derivation from the partial derivative with respect to $\dot{s}$. Though, it does not satisfy the physical features of mass tensors and the matrix $\boldsymbol{M}_{\mathcal{S}}$ is not even quadratic, i.e., $\boldsymbol{M}_{\mathcal{S}} \neq \boldsymbol{M}_{\mathcal{S}}^T$. Finally,

$$-\frac{\partial L}{\partial \dot{s}^b}B_{\mu\,\alpha}^b\dot{r}^\alpha = -M_{\mathcal{S},b\alpha}\,B_{\mu\,\beta}^b\dot{r}^\alpha\,\dot{r}^\beta = -\Upsilon_{\alpha\mu\beta}\,\dot{r}^\alpha\,\dot{r}^\beta\,,\quad \Upsilon_{\alpha\mu\beta} := M_{\mathcal{S},b\alpha}\,B_{\mu\,\beta}^b\,. \tag{4.7}$$

This summarizes the EoMs from (2.78) to

$$\frac{d}{dt}[M_{\mathcal{D},\alpha\mu}\,\dot{r}^\alpha] = \frac{1}{2}\Gamma_{\mathcal{D},\alpha\beta\mu}\,\dot{r}^\alpha\,\dot{r}^\beta - \Upsilon_{\alpha\mu\beta}\,\dot{r}^\alpha\,\dot{r}^\beta + u_\mu\,, \tag{4.8}$$

with u_μ the external generalised forces. Now, including the potential energy U into the Lagrangian $L = T - U$ and defining dU for ease of notation as

$$dU_i := \frac{\partial U}{\partial q^i}\,,\quad dU_\alpha := \frac{\partial U}{\partial r^\alpha}\,,\quad dU_a := \frac{\partial U}{\partial s^a} \tag{4.9}$$

yields

$$U = U(r,s) \;\Rightarrow\; \frac{\partial U}{\partial(r^\alpha,s^a)} = [dU_\alpha\,,\,dU_a]\quad\text{and}\quad \frac{\partial U}{\partial(\dot{r}^\alpha,\dot{s}^a)} = 0 \tag{4.10}$$

$$\Rightarrow\; \frac{\partial L_{\mathcal{D}}}{\partial r^\mu} = \frac{1}{2}\Gamma_{\mathcal{D},\alpha\beta\mu}\,\dot{r}^\alpha\,\dot{r}^\beta - dU_\mu\,,\quad \frac{\partial L_{\mathcal{D}}}{\partial s^a} = -dU_a\,. \tag{4.11}$$

Therefore, the equations of motion (4.8) are extended to

$$\frac{d}{dt}[M_{\mathcal{D},\alpha\mu}\,\dot{r}^\alpha] = \left[\frac{1}{2}\Gamma_{\mathcal{D},\alpha\beta\mu} - \Upsilon_{\alpha\mu\beta}\right]\dot{r}^\alpha\,\dot{r}^\beta - S_\mu^i\,dU_i + u_\mu\,. \tag{4.12}$$

Equation (4.12) can be directly used for simulation and the equivalent pseudo-Hamiltonian equations can be used for controller design which is addressed in Chapter 6.

For deriving the equations of motion for a multi-body mechanical system subject to non-holonomic constraints and consisting of N solid bodies, i.e. for calculating the analytical expressions of the tensors above, the following sequence is proposed that can be automated[2], for instance using the symbolic toolbox of Matlab®:

---

[2] A Matlab-script using the symbolic toolbox to generate equations of motion using these steps was created and is provided under **https://github.com/turnwald/EquationsOfMotion**

1. Define the coordinates q, r and s. Define A.

2. Determine the translational velocities $\boldsymbol{v}_k$ and the angular velocities $\boldsymbol{\Omega}_k$ for every body $k \in 1, \cdots, N$ and, then, derive the translational and rotational kinetic energy for each body, namely T_{vk} and $T_{\Omega k}$ correspondingly.

3. Determine the metric tensors corresponding to each kinetic energy M_{vk} and $M_{\omega k}$. The metric tensor corresponding to each rigid body is then given as $M_k = M_{vk} + M_{\omega k}$.

4. Determine the potential energies U_k for each body.

5. Determine S_α^i, $M_{\mathcal{D}vk}$ and $M_{\mathcal{D}\omega k}$ from (4.2) for each body.

6. Determine $dU_{k,i}$ from (4.9).

7. Determine $\Gamma_{\mathcal{D}k}$ from (4.4) as

$$\Gamma_{\mathcal{D}k,\alpha\beta\mu} = \frac{\partial M_{\mathcal{D},\alpha\beta}}{\partial r^\mu} \ .$$

8. Determine the curvature B, from (2.79).

9. Determine $M_{\mathcal{S}k}$ from (4.6) as

$$M_{\mathcal{S}k,b\alpha} = [0_b^\gamma \ , \ \delta_b^a] \, M_{k,ij} \, S_\alpha^j \ .$$

10. Determine Υ_k from (4.7)
$$\Upsilon_{k,\alpha\mu\beta} = M_{\mathcal{S}k,b\alpha} \, B_{\mu\,\beta}^b \ .$$

A noticeable advantage of this procedure to derive EoMs is its extendibility. For instance, if a new body is added to the system, all of the steps remain and the extended equations are again given by (4.12) where

$$M_{\mathcal{D},\alpha\mu} = \sum_k M_{\mathcal{D}k,\alpha\beta} + M_{\mathcal{D}\omega k,\alpha\beta} \tag{4.13}$$

$$\Gamma_{\mathcal{D},\alpha\beta\mu} = \sum_k \Gamma_{\mathcal{D}k,\alpha\beta\mu} + \Gamma_{\mathcal{D}\omega k,\alpha\beta\mu} \tag{4.14}$$

$$M_{\mathcal{S},b\alpha} = \sum_k M_{\mathcal{S}k,b\alpha} \Rightarrow \Upsilon_{\alpha\mu\beta} = \sum_k \Upsilon_{k,\alpha\mu\beta} \tag{4.15}$$

$$dU_i = \sum_k dU_{k,i} \ . \tag{4.16}$$

Note that (4.12) is an ODE and, as stated in [KM97], is equivalent to the (pseudo-)Hamiltonian equations (2.86) that can be used, for instance for designing passivity-based controllers and observers. The only missing term from (2.86) is R that is determined by

$$R_{\mu\beta} = -\Upsilon_{\alpha\mu\beta} \, \dot{r}^\alpha \big|_{\dot{r}^\alpha = M^{\mathcal{D},\alpha\beta} \rho_\alpha} \ . \tag{4.17}$$

To obtain the EoMs in the well-known form, e.g. from the robotics literature, the left hand-side of (4.12) needs to be rewritten as

$$\frac{d}{dt}[M_{\mathcal{D},\alpha\mu}\,\dot{r}^{\alpha}] = M_{\mathcal{D},\alpha\mu}\,\ddot{r}^{\alpha} + \frac{\partial M_{\mathcal{D},\alpha\mu}}{\partial r^{\beta}}\,\dot{r}^{\mu}\dot{r}^{\beta} = M_{\mathcal{D},\alpha\mu}\,\ddot{r}^{\alpha} + \Gamma_{\mathcal{D},\alpha\mu\beta}\,\dot{r}^{\alpha}\dot{r}^{\beta} \tag{4.18}$$

and therefore

$$M_{\mathcal{D},\alpha\mu}\,\ddot{r}^{\alpha} = \left[\frac{1}{2}\Gamma_{\mathcal{D},\alpha\beta\mu} - \Gamma_{\mathcal{D},\alpha\mu\beta} - \Upsilon_{\alpha\mu\beta}\right]\dot{r}^{\alpha}\dot{r}^{\beta} - S^{i}_{\mu}\,dU_{i} + \tau_{\mu} \tag{4.19}$$

$$\Rightarrow\ M_{\mathcal{D},\alpha\mu}\,\ddot{r}^{\alpha} = \Lambda_{\mathcal{D},\alpha\beta\mu}\,\dot{r}^{\alpha}\dot{r}^{\beta} - S^{i}_{\mu}\,dU_{i} + \tau_{\mu}\,. \tag{4.20}$$

with the tensor $\Lambda_{\mathcal{D}}$ being defined as

$$\Lambda_{\mathcal{D},\alpha\beta\mu} := \frac{1}{2}\Gamma_{\mathcal{D},\alpha\beta\mu} - \Gamma_{\mathcal{D},\alpha\mu\beta} - \Upsilon_{\alpha\mu\beta} \tag{4.21}$$

(4.12) and (4.20) are the general EoMs of the vehicle which are used in simulations throughout this work. In fact, the outcome of the simulation has to be the same. Numerically, it is better to use (4.12) for simulating the system behaviour because of the extra term in (4.20). For trajectory planing however, (4.20) is suggested because the second derivatives can be accessed directly and the inverse of the mass tensor is not required.

4.3 Derivation of a planar single-track vehicle model

Figure 4.1 illustrates a single-track vehicle model that simplifies the planar motion of a four-wheeled vehicle. However, it can also represent a simplified model for a bicycle ignoring the leaning motion. Note that the schematics in Figure 4.1 illustrates also a top view of the TWV considered in the next section, however with an exaggeratedly large trail Δ for a better visualisation of the constellation. It is mainly chosen because of brief expressions in the equations of motion, which is not the case later for more detailed bicycle model. One may interpret this model as the projection or the *shadow* of the TWV on the ground, or, a bicycle with support wheels on the ground.

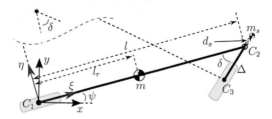

Figure 4.1: Schematics of a planar single-track vehicle with positive trail

The contact point of the rear wheel C_1 is the origin of the body-fixed frame with the coordinates ξ and η, which is rotated by the yaw angle ψ regarding the world frame with the coordinates x and y. Note that the world coordinates are not attached to the vehicle and only drawn at C_1 for a better comparison. The contact point of the front wheel is C_3. The point C_2 is the steering point that is the intersection of the extension of the steering axis with the ground plane. The positive trail Δ measures the distance between the points C_2 and C_3. The wheel base l is the length between C_1 and C_2 and m is the total mass of the rear frame. The distance between the Centre Of Mass (CoM) and the rear contact point C_1 is l_r. The mass of the steering mechanism is m_s, which is assumed to be shifted by d_s in front of the steering axis. The steering angle is δ. Note that δ is not the angel of the steering bar, but the so-called *real steering angle*. This is basically the projection of the steering angel on the ground and what is responsible for actual steering of the vehicle. The torque acting on the steering mechanism is denoted by u_δ and the force acting on the rear wheel to move the vehicle forward along the ξ-axis is denoted by u_ξ. By steering, i.e. changing the angle δ , the CoM of the rear frame is shifted laterally with regard to the contact point line $C_1 - C_3$, as show in Figure 4.1. This is maybe the most important reasons why a standing (not moving) bicycle can be balanced by steering only.

Using the scheme proposed above to derive the EoMs, the generalised coordinates need to be chosen, first. The vehicle consists of two rigid bodies, the rear frame and the steering assembly, and has four degrees of freedom. The position of the vehicle and its orientation as well as the steering angle can be chosen as generalised coordinates q. In fact, these entities, in addition to the leaning angle, are chosen to be the generalised coordinates of the bicycle model in the next section. For a planar *Naive* vehicle model ($\Delta = d_s = 0$), however, there exists a better choice of the generalised coordinates that reduces the size and the complexity of the resulting EoMs noticeably. That is, inspired by [GM95], using the so-called steering variable $\sigma := \frac{\tan \delta}{l}$ instead of the steering angle δ. Since the objective of this section is to demonstrate the steps above by a minimal example, the EoMs and the corresponding tensors are derived below for the *Naive* planar vehicle. Of course, some of the steps are trivial or even unnecessary for simple mechanical systems. Yet, they are included in an automated generation of EoMs. Furthermore, the required terms for the Hamiltonian formulation of the same equations are determined.

Step 1: The generalised coordinates are chosen to be

$$q = \begin{bmatrix} \xi & \sigma & \eta & \psi \end{bmatrix}^T .$$

The vehicle is subject to two nonholonomic constraints, that are, the lateral velocity of each idealised wheel is zero. For the rear wheel, this means that $\dot{\eta} = 0$. For the front wheel, this implies geometrically that

$$\dot{\psi} = \dot{\xi} \frac{\tan \delta}{l} = \dot{\xi} \sigma .$$

Therefore, the base coordinates r, the fiber coordinates s and the corresponding Ehresmann

A are given as

$$\dot{s} = \begin{bmatrix} \dot{\eta} \\ \dot{\psi} \end{bmatrix} \ , \quad \dot{r} = \begin{bmatrix} \dot{\xi} \\ \dot{\sigma} \end{bmatrix} \ , \quad A(r) = \begin{bmatrix} 0 & 0 \\ -\sigma & 0 \end{bmatrix}. \tag{4.22}$$

Step 2: The translational velocity of the CoM of the rear frame v_c and its rotational velocity ω_c as well as the corresponding kinetic energies are obtained as

$$v_c = \begin{bmatrix} \dot{\xi} \\ \dot{\eta} + l_r \dot{\psi} \end{bmatrix} \ , \quad \omega_c = \dot{\psi} \ , \tag{4.23}$$

$$T_{vc} = \frac{1}{2} m \dot{\xi}^2 + \frac{1}{2} m (\dot{\eta} + l_r \dot{\psi})^2 \ , \quad T_{\Omega c} = \frac{1}{2} J \dot{\psi}^2 \ , \tag{4.24}$$

with J the inertia of the rear frame.

The velocity of the CoM of the steering assembly v_s and its rotational velocity Ω_s as well as the corresponding kinetic energies are obtained as

$$v_s = \begin{bmatrix} \dot{\xi} \\ \dot{\eta} + l \dot{\psi} \end{bmatrix} \ , \quad \Omega_s = \dot{\delta} \tag{4.25}$$

$$T_{vs} = \frac{1}{2} m_s \dot{\xi}^2 + \frac{1}{2} m_s (\dot{\eta} + l \dot{\psi})^2 \ , \quad T_{\Omega s} = \frac{1}{2} J_s \dot{\delta}^2 = \frac{1}{2} J_\sigma \dot{\sigma}^2 \ , \quad J_\sigma(\sigma) = \frac{J_s \, l^2}{(1 + l \, \sigma)^2} \ , \tag{4.26}$$

with J_s the inertia of the steering mechanism.

Step 3: The metric tensors are determined as

$$M_{vc} = \begin{bmatrix} m & 0 & 0 & 0 \\ 0 & 0 & 0 & 0 \\ 0 & 0 & m & m \, l_r \\ 0 & 0 & m \, l_r & m \, l_r^2 \end{bmatrix} \ , \quad M_{wc} = \begin{bmatrix} 0 & 0 & 0 & 0 \\ 0 & 0 & 0 & 0 \\ 0 & 0 & 0 & 0 \\ 0 & 0 & 0 & J \end{bmatrix} \tag{4.27}$$

$$M_{vs} = \begin{bmatrix} m_s & 0 & 0 & 0 \\ 0 & 0 & 0 & 0 \\ 0 & 0 & m_s & m_s \, l \\ 0 & 0 & m_s \, l & m_s \, l^2 \end{bmatrix} \ , \quad M_{ws} = \begin{bmatrix} 0 & 0 & 0 & 0 \\ 0 & J_\sigma(\sigma) & 0 & 0 \\ 0 & 0 & 0 & 0 \\ 0 & 0 & 0 & 0 \end{bmatrix} \tag{4.28}$$

Step 4: The potential energy of this vehicle is zero $U = 0$.

Step 5: To obtain the reduces metric tensors, first S is required that is given by

$$S = \begin{bmatrix} 1 & 0 & 0 & 0 \\ 0 & 1 & \sigma & 0 \end{bmatrix} \ . \tag{4.29}$$

Therefore,

$$M_{Dc} = \begin{bmatrix} J \sigma^2 + m & 0 \\ 0 & 0 \end{bmatrix} \ , \quad M_{Ds} = \begin{bmatrix} m_s & 0 \\ 0 & J_\sigma(\sigma) \end{bmatrix} \ , \tag{4.30}$$

which leads to the constrained Lagrangian

$$L_D = \frac{1}{2} \dot{r}^T M_D \, \dot{r} \ , \quad M_D = \begin{bmatrix} J \sigma^2 + m + m_s & 0 \\ 0 & J_\sigma(\sigma) \end{bmatrix} \ . \tag{4.31}$$

Step 6: Consequently, $dU = 0$.

Steps 7,8: The three dimensional tensors Γ and B are calculated by differentiating $\boldsymbol{M}_{\mathcal{D}}$ and A respectively.

Step 9: The tensor $M_{\mathcal{S}c}$ and $M_{\mathcal{S}s}$ are

$$M_{\mathcal{S}c} = \begin{bmatrix} m\,l_r\,\sigma & 0 \\ (m\,l_r^2 + J)\,\sigma & 0 \end{bmatrix}, \quad M_{\mathcal{S}s} = \begin{bmatrix} m_s\,l\,\sigma & 0 \\ m_s\,l^2\,\sigma & 0 \end{bmatrix}. \tag{4.32}$$

Step 10: The three dimensional tensor Υ is determined using $M_{\mathcal{S}}$ and B.

Finally, the EoMs for the *Naive* single-track model on the constraint space are given by (4.20) using matrix notation as

$$\boldsymbol{M}_{\mathcal{D}}(\boldsymbol{r})\,\ddot{\boldsymbol{r}} + \begin{bmatrix} J\,\sigma\,\dot{\sigma}\,\dot{\xi} \\ -\frac{J_S l^3\,\dot{\sigma}^2}{(l\,\sigma+1)^3} \end{bmatrix} = \begin{bmatrix} u_\xi \\ u_\sigma \end{bmatrix}, \tag{4.33}$$

where u_σ is the torque on the steering variable and u_ξ the force applied to the rear wheel.

In order to derive the EoMs in the pseudo-Hamiltonian form, the transformed impulse coordinates on are defined by (2.81) as

$$\boldsymbol{\rho} = \begin{bmatrix} m\dot{\xi} + \sigma\,(J + m\,l_r^2)\,\dot{\psi} + \sigma\,l_r\,m\dot{\eta} \\ J_\sigma\,(\sigma)\,\dot{\sigma} \end{bmatrix}. \tag{4.34}$$

The pseudo-Hamiltonian system is given by (2.86) where

$$\boldsymbol{S} = \begin{bmatrix} 1 & 0 & 0 & 0 \\ 0 & 1 & \sigma & 0 \end{bmatrix}^T, \quad \boldsymbol{R} = \begin{bmatrix} 0 & \frac{J\,\sigma\,\rho_1}{J\,\sigma^2+m} \\ -\frac{J\,\sigma\,\rho_1}{J\,\sigma^2+m} & 0 \end{bmatrix}$$

$$H_{\mathcal{M}} = \frac{1}{2}\frac{\rho_1^2}{J\,\sigma^2 + m} + \frac{1}{2}\frac{\rho_2^2}{J_\sigma(\sigma)}. \tag{4.35}$$

For sake of completeness, the impulse coordinates corresponding to the vertical space are also calculated by

$$\check{\boldsymbol{\rho}} = \begin{bmatrix} l_r m \\ (J + m\,l_r^2) \end{bmatrix} \cdot \frac{\sigma^2\,(J + m L_r^2)\,\dot{\psi} + \sigma^2\,l_r m\dot{\eta}}{(J + m\,l_r^2)\,\sigma^2 + m} \tag{4.36}$$

which are, however, not used in the controller design nor in the simulation.

Note that the world coordinates are given by

$$\dot{x} = \dot{\xi}\,\cos\psi\,, \quad \dot{y} = \dot{\xi}\,\sin\psi\,. \tag{4.37}$$

4.4 Derivation of the bicycle model: A 2-body approach

The bicycle model derived in this section can be considered as an extension of the model of the planar vehicle from the previous section. Since the steering variable cannot be used for

$\Delta \neq 0$, the terms in the tensors are no more compact enough to be illustrated. Therefore, the restyling tensors are omitted here to save space [3].

4.4.1 Kinematics and the generalised coordinates

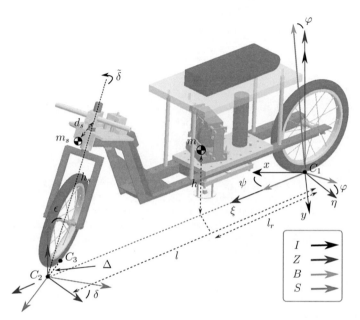

Figure 4.2: Schematics of a two-wheeled vehicle including coordinate systems and parameters

Figure 4.2 shows the schematics of the prototype bicycle developed at the institute for control systems which is explained more in Chapter 7. The total mass of the rear frame including a passive rider is m. The total mass of the steering mechanism including the handlebar and the front wheel is m_s. The wheelbase is l and Δ denotes the trail on the ground. The CoM of the rear frame is given by h and l_r while the CoM of the steering mechanism is shifted by d_s and h_s with respect to the steering axis. The caster angle is ϵ . The model parameters are summarised in Table 4.2.

---

[3]The analytically calculated tensors can be found on GitHub: `https://github.com/turnwald/BikeModel`. One may downlaod the entire Simulink-Project to use the internal SVN of Simulink or, alternatively, only the tensors as Matlab-functions in a zip-file.

Notation	Origin	Axes	Description
I	C_1	$[x, y, *]$	World coordinate system[4]
Z	C_1	$[\xi, \eta, *]$	I rotated by ψ about z, local coordinate system
B	C_1	$[\xi, *, *]^{\text{a}}$	Z rotated by φ about ξ, attached to the rear frame
$\tilde{S}$	C_2	$[*, *, *]$	B rotated by $\tilde{\delta}$, attached to the steering mechanism
S	C_2	$[*, *, *]$	Z rotated by δ, projection of $\tilde{S}$ on the ground

[a]Axes that are not explicitly used or that no letter is assigned to are noted by $*$ to reduce the number of used letters.

Table 4.1: Coordinate systems used for the model derivation of the bicycle

The steering angel is $\tilde{\delta}$. This, however, does not directly determine the motion of the bicycle on a curve. In most literature, therefore, the so-called *real steering angel* δ is taken to describe the kinematics of the bicycle. That is basically the projection of the steering angle on the ground plane and the following relation holds ([Cos06]):

$$\cos\varphi \tan\delta = \cos\varepsilon \tan\tilde{\delta}. \qquad (4.38)$$

Using the real steering angle as a generalised coordinate simplifies the kinematics of the vehicle. Similar to the planar vehicle from the previous section, the rear wheel contact point C_1 is considered to define the bicycle position, that is given by (ξ, η) in the body-fixed CS, Z. The orientation angle of the rear wheel in the world CS, ψ, and the leaning angle φ complete the required generalised coordinates. In summary, the generalised coordinates are chosen as

$$\boldsymbol{q} = \begin{bmatrix} \xi, & \varphi, & \delta, & \psi, & \eta \end{bmatrix}^T. \qquad (4.39)$$

4.4.2 Kinetic energy

Below, sub-indexes $*_c$ are used for the bicycle frame and $*_s$ for the steering mechanism. To determine the kinetic energy, the velocities and angular velocities of each body are required. The coordinate systems used in this section are summarised in Table 4.1. The corresponding rotation matrices are given by

$$^{ZB}\boldsymbol{R} = Rot_1(\varphi) \,, \quad ^{B\tilde{S}}\boldsymbol{R} = Rot_2(-\epsilon)\,Rot_3(\tilde{\delta}) \qquad (4.40)$$

The angular velocity of Z is

$$^{IZ}_{Z}\boldsymbol{\omega} = \begin{pmatrix} 0 & 0 & \dot{\psi} \end{pmatrix}^T. \qquad (4.41)$$

In the model from [GM95], a very important entity, the trail, is missing. Beside the self-stability, it also allows the bicycle to be balanced by steering at zero velocity. In [ZLYS11], this model is modified by adding an estimated height change for the CoM. This makes the potential energy depend on a mixed term of φ and δ. This shift is also explained in [Cos06].

However, it also affects the kinematics and, thus, the kinetic energy of the system.
Here, the shift of the CoM due to steering is already considered in the position vector as the last term in (4.42) below. In this way, not only the potential energy is corrected, but also the kinematic interconnection of the steering and the leaning is described more accurately. The CoM of the rear frame is given by the position vector $\boldsymbol{w}_c$ and the corresponding velocity $\boldsymbol{v}_c$ as

$$_Z\boldsymbol{w}_c = \begin{pmatrix} l_r \\ 0 \\ 0 \end{pmatrix} + {}^{ZB}\boldsymbol{R} \begin{bmatrix} 0 \\ 0 \\ h \end{bmatrix} + \begin{bmatrix} 0 \\ \frac{l_r \Delta}{l} \cos \epsilon \sin \delta \sin \varphi \\ 0 \end{bmatrix} \tag{4.42}$$

$$_Z\boldsymbol{v}_c = \begin{bmatrix} \dot{\xi} \\ \dot{\eta} \\ 0 \end{bmatrix} + {}^{IZ}_Z\boldsymbol{\omega} \times {}_Z\boldsymbol{w}_c + \frac{\partial^T {}_Z\boldsymbol{w}_c}{\partial \boldsymbol{q}} \dot{\boldsymbol{q}} . \tag{4.43}$$

The position of the CoM of the steering mechanism is

$$_Z\boldsymbol{w}_s = \begin{bmatrix} l \\ 0 \\ 0 \end{bmatrix} + {}^{ZB}\boldsymbol{R} \, {}^{B\tilde{S}}\boldsymbol{R} \begin{bmatrix} d_s \\ 0 \\ h_s \end{bmatrix} . \tag{4.44}$$

Note that h_s is along the steering axis and not perpendicular to the ground plane. Because of the choice of the generalised coordinates, all equations need to be rewritten in terms of δ to eliminate $\tilde{\delta}$. The relation (4.38) can be approximated as follows:

$$\cos\epsilon \tan \tilde{\delta} = \cos\varphi \tan \delta \Rightarrow \begin{cases} \sin\tilde{\delta} \simeq \frac{\cos\varphi}{\cos\epsilon} \sin \delta \\ \cos\tilde{\delta} \simeq \cos \delta \end{cases} . \tag{4.45}$$

After the transformation of the position vector, the corresponding velocity is calculated similar to (4.43). Further, one needs to transform the velocity $\dot{\tilde{\delta}}$ into S. The time derivative of (4.38) and substitution yield

$$-\dot{\varphi} \sin\varphi \tan \delta + \dot{\delta}(1 + \tan^2 \delta) \cos\varphi = \dot{\tilde{\delta}} \cos\epsilon \left(1 + \frac{\cos\varphi^2}{\cos\epsilon^2} \tan^2 \delta \right) . \tag{4.46}$$

For calculating the translational kinetic energy of the steering T_{vs}, it is simpler to first use the coordinates of $\tilde{S}$ and then use the Jacobian to transform the metric tensor. The metric tensor of T_{vs} with regard to $\tilde{\delta}$ is given as $\tilde{M}_{vs,ij}$ and the relation

$$M_{s,ij} = \tilde{M}_{s,ij} \, Q_{ij} \, Q_{ji} \Big|_{\tilde{q}=\Phi^{-1}(q)} \tag{4.47}$$

$$Q_{ij} = \begin{bmatrix} 1 & 0 & 0 & 0 & 0 \\ 0 & \frac{\cos\varphi \left(\tan\delta^2 + 1 \right)}{\cos\varphi^2 \tan\delta^2 + 1} & -\frac{\sin\varphi \tan\delta}{\cos\varphi^2 \tan\delta^2 + 1} & 0 & 0 \\ 0 & 0 & 1 & 0 & 0 \\ 0 & 0 & 0 & 1 & 0 \\ 0 & 0 & 0 & 0 & 1 \end{bmatrix} \tag{4.48}$$

holds, where Q_{ij} are the components of the Jacobian between S and $\tilde{S}$ and is $\tilde{q} = \Phi(q)$ the transformation to S. Recall that the transformation Φ is given by (4.38). In this case, however, not the actual inverse transformation, but the approximated inverse transformation from (4.45) is used. Q is derived from (4.46).

The angular velocities are each given in the body-fixed coordinate systems to keep the inertia tensors simple. It is assumed that coordinate systems B and $\tilde{S}$ correspond to the main inertia axes of the rear frame and the steering mechanism accordingly. The rotation of rear frame is given by

$$
{}_B\Omega_c = \begin{bmatrix} \dot{\varphi} \\ 0 \\ 0 \end{bmatrix} + {}^{ZB}R^T \begin{bmatrix} 0 \\ 0 \\ \dot{\psi} \end{bmatrix} = \begin{bmatrix} \dot{\varphi} \\ \dot{\psi}\,\sin\varphi \\ \dot{\psi}\,\cos\varphi \end{bmatrix} \tag{4.49}
$$

The rotation of the steering mechanism is given by

$$
{}_{\tilde{S}}\Omega_s = {}^{B\tilde{S}}R^T \, {}_B\Omega_c + \begin{bmatrix} 0 \\ 0 \\ \dot{\tilde{\delta}} \end{bmatrix} + \begin{bmatrix} 0 \\ \frac{\dot{\xi}}{R_w} \\ 0 \end{bmatrix} \tag{4.50}
$$

with R_w being the wheel radius of the front wheel. This needs to be transformed, too, using (4.46) and (4.45). Note that (4.50) basically consists of two parts. The first part corresponds to the rotation of the front assembly and generates the kinetic energy of the steering. The last term corresponds to the gyroscopic force due to the rotation of the front wheel. The gyroscopic force of the front wheel has a non-negligible effect on self-alignment, although it is known that a self-stabilisation is also possible for a bicycle without it.

Assuming no pitch, the total kinetic energy is given by

$$
T = T_{vc} + T_{vs} + T_{\Omega c} + T_{\Omega s} \tag{4.51}
$$
$$
T_{vc} = \frac{1}{2}\,m\,\langle v_c \cdot v_c \rangle = \frac{1}{2}\,\dot{q}^i\,\dot{q}^j\,M_{vc,\,ij}\,, \quad T_{vs} = \frac{1}{2}\,m_s\,\langle v_s \cdot v_s \rangle = \frac{1}{2}\,\dot{q}^i\,\dot{q}^j\,M_{vs,\,ij}
$$
$$
T_{\Omega c} = \frac{1}{2}\,\langle \Omega_c \cdot J_c\,\Omega_c \rangle = \frac{1}{2}\,\dot{q}^i\,\dot{q}^j\,M_{\omega c,\,ij}\,, \quad T_{\Omega s} = \frac{1}{2}\,\langle \Omega_s \cdot J_s\,\Omega_s \rangle = \frac{1}{2}\,\dot{q}^i\,\dot{q}^j\,M_{\omega s,\,ij}
$$

with $\langle \cdot \rangle$ the scalar product on the configuration space. $J_c = \mathrm{diag}\,(J_x, *, J_z)$ and $J_s = \mathrm{diag}\,(J_{sx}, J_w, J_s)$ are the inertia matrices of the rear frame and the steering mechanism accordingly, where J_w is the inertia of the front wheel.

4.4.3 The gravity and the potential energy

The effect of the gravity creates a balance contradicting the centrifugal force applied on the bicycle CoM in a curve ride. Enabling a non-zero leaning angle, it leads to the bicycle being steerable as stated in [DGG+17]. Also, it is a part of the self-stabilisation mechanism by exerting a force on the steering CoM. It is worth noting that the gradient of the potential energy causes a torque about the axis $C_1 - C_3$. In other words, the rotation of the rear frame, when the bicycle falls over, is about the axis $C_1 - C_3$ and not the ξ-axis. Since the shift of the CoM is already defined in (4.42), the potential energy is simply given by the 3rd component of each position vector as $U = U_c + U_s$.

$$U_c = m\,g \begin{bmatrix} 0 & 0 & 1 \end{bmatrix} {}_Z \boldsymbol{w}_c \,, \quad U_s = m_s\,g \begin{bmatrix} 0 & 0 & 1 \end{bmatrix} {}_Z \boldsymbol{w}_s \tag{4.52}$$

Note that the resulting position vector leading to U_s in (4.52) is consistent with the proposed height change due to steering action by [ZLYS11].

4.4.4 Nonholonomic constraints

Assuming no slip, the nonholonomic constraints state that the lateral velocity of each wheel is zero. For the rear wheel, this means $\dot{\eta} = 0$. For the front wheel the lateral velocity of the point C_3 is zero. To calculate that, the position vector of C_3 is required that is

$$_Z\boldsymbol{w}_{C_3} = \begin{bmatrix} l - \Delta \cos\delta \\ -\Delta \sin\delta \\ 0 \end{bmatrix} . \tag{4.53}$$

Therefore, the velocity of the point C_3 is

$$_Z\boldsymbol{v}_{C_3} = \begin{bmatrix} \dot{\xi} \\ \dot{\eta} \\ 0 \end{bmatrix} + {}_Z^{IZ}\boldsymbol{\omega} \times {}_Z\boldsymbol{w}_{C_3} + \frac{\partial^T {}_Z\boldsymbol{w}_{C_3}}{\partial q} \dot{\boldsymbol{q}}$$

$$= \begin{bmatrix} \dot{\xi} \\ \dot{\eta} \\ 0 \end{bmatrix} + \begin{bmatrix} 0 \\ 0 \\ \dot{\psi} \end{bmatrix} \times \begin{bmatrix} l - \Delta \cos\delta \\ -\Delta \sin\delta \\ 0 \end{bmatrix} + \begin{bmatrix} \dot{\delta}\,\Delta \sin\delta \\ -\dot{\delta}\,\Delta \cos\delta \\ 0 \end{bmatrix} \tag{4.54}$$

witch results in

$$_Z\boldsymbol{v}_{C_3} = \begin{bmatrix} \dot{\xi} + \dot{\delta}\,\Delta \sin\delta + \dot{\psi}\,\Delta \sin\delta \\ \dot{\eta} - \dot{\delta}\,\Delta \cos\delta - \dot{\psi}\,\Delta \cos\delta + l\dot{\psi} \\ 0 \end{bmatrix} . \tag{4.55}$$

The lateral velocity of C_3 is the second component of ${}_S\boldsymbol{v}_{C_3}$ that is required to be zero. This is determined by

$$\begin{bmatrix} 0 & 1 & 0 \end{bmatrix} {}_S\boldsymbol{v}_{C_3} = \begin{bmatrix} 0 & 1 & 0 \end{bmatrix} {}^{ZS}\boldsymbol{R}^{-1} {}_Z\boldsymbol{v}_{C_3} = 0 \tag{4.56}$$

that is, noticing $\dot{\eta} = 0$,

$$0 = \begin{bmatrix} 0 & 1 & 0 \end{bmatrix} \begin{bmatrix} \cos\delta & \sin\delta & 0 \\ -\sin\delta & \cos\delta & 0 \\ 0 & 0 & 1 \end{bmatrix} {}_Z\boldsymbol{v}_{C_3}$$

$$= \begin{bmatrix} -\sin\delta & \cos\delta & 0 \end{bmatrix} \begin{bmatrix} \dot{\xi} + \dot{\delta}\,\Delta \sin\delta + \dot{\psi}\,\Delta \sin\delta \\ \dot{\eta} - \dot{\delta}\,\Delta \cos\delta - \dot{\psi}\,\Delta \cos\delta + l\dot{\psi} \\ 0 \end{bmatrix}$$

$$= -(\dot{\delta} + \dot{\psi})\,\Delta \sin^2\delta - \dot{\delta} \sin\delta - (\dot{\delta} + \dot{\psi})\,\Delta \cos^2\delta + l\,\dot{\psi} \cos\delta \tag{4.57}$$

which results in

$$\dot{\psi}\left(l\cos\delta - \Delta\right) = \dot{\delta}\,\Delta + \dot{\xi}\,\sin\delta \tag{4.58}$$

or

$$\dot{\psi} = \frac{\sin\delta}{l\cos\delta - \Delta}\,\dot{\xi} + \frac{\Delta}{l\cos\delta - \Delta}\,\dot{\delta}\,. \tag{4.59}$$

Choosing

$$\boldsymbol{r} = \begin{bmatrix} \xi \\ \varphi \\ \delta \end{bmatrix},\ \boldsymbol{s} = \begin{bmatrix} \psi \\ \eta \end{bmatrix}, \tag{4.60}$$

this leads to the constraint equations

$$\dot{\boldsymbol{s}} = -\boldsymbol{A}\dot{\boldsymbol{r}} \ \Rightarrow\ \begin{bmatrix} \dot{\psi} \\ \dot{\eta} \end{bmatrix} = -\begin{bmatrix} \frac{-\sin\delta}{l\cos\delta - \Delta} & 0 & \frac{-\Delta}{l\cos\delta - \Delta} \\ 0 & 0 & 0 \end{bmatrix} \begin{bmatrix} \dot{\xi} \\ \dot{\varphi} \\ \dot{\delta} \end{bmatrix}. \tag{4.61}$$

In conclusion, the equations of motion of the vehicle are determined by (4.12) or (4.20) for simulations and model validation with the vector of generalised forces $\boldsymbol{u} = \begin{bmatrix} u_\xi & u_\varphi & u_\delta \end{bmatrix}^T$, which contains

- u_ξ: The force applied on the rear wheel. Note that this force is in general created by a toque rotating the rear wheel.

- u_φ: The torque on the rear frame that causes the bicycle to lean to the side. This torque does not exist on regular bicycles or motorcycles, although there are already approaches to use special actuators to create such a torque. In the simulations below, this input is mainly used as a source for external disturbance.

- u_δ: The torque on the steering angle.

4.5 Simulations and model validation

For model validation, different scenarios are considered and simulation results are analysed regarding physically consistent expected behaviour. Also, a comparison to the well-known benchmark model is presented. Furthermore, a prototype is developed to experimentally investigate the validity of the model that is explained in more detail in Chapter 8.

4.5.1 Numeric simulations of different scenarios

The following scenarios are chosen to each represent a relevant effect. The model parameters from Table 4.2 are used which are chosen to comply with those from [TWB10]. Note that

Table 4.2: Parameters used for the numerical simulation (all in SI).

Parameter	Value	Short description
$\mathrm{diag}\left(J_x, *, J\right)$	$\mathrm{diag}\left(10.3, *, 13.5\right)$	Inertia of rear frame
$\mathrm{diag}\left(J_{sx}, *, J_s\right)$	$\mathrm{diag}\left(3.8, *, 0.08\right)$	Inertia of steering mechanism
$\left(h, 0, l_r\right)$	$\left(0.8, 0, 0.5\right)$	Centre of mass of rear frame
$\left(h_s, 0, d_s\right)$	$\left(0.6, 0, 0.15\right)$	Centre of mass of steering mechanism
m	50	Mass of the rear frame
m_s	8	Mass of the steering mechanism
J_w	0.15	Inertia of front wheel
Δ	0.09	Trail
R_w	0.3	Radius of front wheel
g	9.81	Gravity constant
l	1.2	Wheelbase
ϵ	$19\,^\circ$	Head Angle

the parameters are not identical due to different model settings. Note that for sake of better readability in what fallows the forward velocity of the vehicle is denoted by v that is $v = \dot{\xi}$.

1. **Balance by steering for $v = 0\,\mathrm{m/s}$**: Due to steering, the CoM of the bicycle is shifted and lays no more above the line connecting the wheel contact points. This makes the uncontrolled bicycle fall over, can also be used to balance a bicycle by only steering at zero velocity. To verify this fact, a steering torque is applied as a function of the steering and leaning angles. The the initial leaning angle is to $\varphi(0) = 1°$. To avoid accelerating the vehicle due to steering, also a negative force was added to the rear wheel, that is equivalent to a break. The acceleration is caused by the constraint forces from the ground on the front wheel. In fact, this is how Snake-Boards ([KM97]) move forward. The results are shown in Figure 4.3. One can see that the leaning angle φ and the orientation angel ψ are affected by the steering although the vehicle is almost steady. Note that the objective of this simulation is not to successfully apply a balance control, but rather to illustrate the physical possibility of doing so.

2. **Accelerating from $v = 0\,\mathrm{m/s}$**: Because of the positive trail, accelerating leads to a self-alignment of the steering angle for $\delta \neq 0$. To investigate this fact, the bicycle is accelerated from $v(0) = 0\,\mathrm{m/s}$ up to $v(t_{end}) = 5.1\,\mathrm{m/s}$ with the initial steering angle set to $\delta(t = 0) = 5°$. If the acceleration is large enough, the self-alignment makes the steering angle converge to zero and the bicycle moves forward without falling over. For generating the force on the rear wheel $u_\xi = -85\left(\dot{\xi} - 5.1\,\mathrm{m/s}\right)$ is used that creates the results from Figure 4.4. Interestingly, due to the same fact, the wheels of a trolley move to the behind once it is pushed forward. Also, this is among the reasons why a steering wheel of a car turns back to the zero position by accelerating after a curve.

3. **Torque-impulse on the steering bar**: The input-output behaviour of a bicycle is non-minimum-phase from steering torque u_δ to the orientation ψ. As explained in Section 3.1, in order to start a left-curve, the rider has to steer right, first. To illustrate

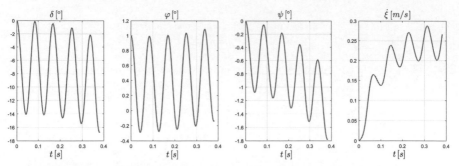

Figure 4.3: Simulation results for Scenario 1: Balance by steering for $v = 0\,\mathrm{m/s}$

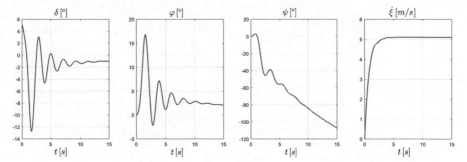

Figure 4.4: Simulation results for Scenario 2: Accelerating from $v = 0\,\mathrm{m/s}$

this fact, a short simulation of the uncontrolled bicycle is run with an initial forward velocity $v = 3.7\,\mathrm{m/s}$. At $t = 3\,\mathrm{s}$, an impulse is applied to the input u_δ, which yields the results from Figure 4.5. The non-minimum-phase behaviour is clearly visible in the graph of δ as marked with a red circle.

4. **Side-kick for** $v > v_w$: An uncontrolled bicycle moving with a higher forward velocity than the so-called *weave velocity* v_w is self-stable, as explained in Section 3.2. To verify this, an impulse is applied to the input u_φ, which is equivalent to a *side-kick* on the vehicle body. The simulation results for $v = 3.7\,\mathrm{m/s}$ are illustrated in Figure 4.6. One may observe that the uncontrolled vehicle is stabilised, i.e., the leaning angle φ and the steering angel δ converge to zero after a while. Note that the orientation angle ψ is changed and remains at a different value other that zero. This is a well-known behaviour of bicycles that is explained by the nonholonomic constraints.

5. **Side-kick for** $v > v_c$: A bicycle moving faster than the so-called *capsize velocity* it is no more self-stable, as explained in Section 3.2. This is because the self-alignment torque is too high and the steering angel converges to zero too fast. Thus, the bicycle behaves

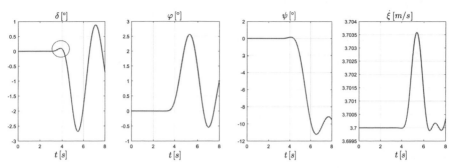

Figure 4.5: Simulation results for Scenario 3: Non-minimum-phase behaviour

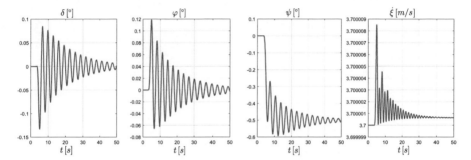

Figure 4.6: Simulation results for Scenario 4: Side-kick for $v > v_w$

similar to the case in which steering is locked and falls over due to a disturbance. This fact is illustrated by running the simulation of the uncontrolled bicycle at $v = 5.5\,\mathrm{m/s}$ subject to a side-kick. The results are shown in Figure 4.7.

6. **Control of φ by u_δ only**: To ride a bicycle on a curve, one could design a controller to maintain a constant leaning angle. Using the torque on the steering bar as input, a simple state feedback is applied to settle the desired leaning angle $\varphi^* = 8°$. The feedback $u_\delta = -20\,(\varphi - \varphi^*) - 30\,\dot{\varphi}$ is used in the simulations to create the results from Figure 4.8. Note that the feedback law above is not designed specifically to track the desired angel with no error, but rather the possibility of doing so using the steering torque only.

7. **Control of φ by u_φ only**: Using the torque on the leaning angle as an input, a simple state feed-back may settle a constant desired lean angle φ^*. One can think of this as riding the bicycle hands-free and moving only the body. The Scenarios 6 and 7 verify the rideability of the bicycle by two independent inputs. The feedback $u_\varphi = -50\,\varphi - 30\,\dot{\varphi}$ is used to create the results from Figure 4.9.

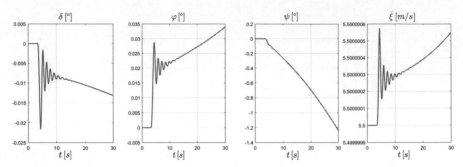

Figure 4.7: Simulation results for Scenario 5: Side-kick for $v > v_c$

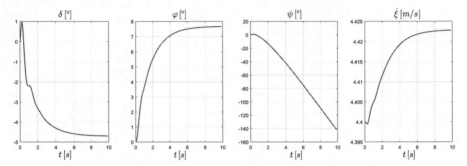

Figure 4.8: Simulation results for Scenario 6: Control of φ by only u_δ

4.5.2 Constant velocity and comparison to the benchmark model

In this section, the proposed model from Section 4.4 validated by comparing it to the so-called benchmark model. Note that in this section, the indexes ($\alpha', \beta', \mu' \in \{2,3\}$) are used to denote the last two elements of the base coordinates r. The benchmark model considered the leaning angle φ and the steering angel δ as well as the corresponding rates $\dot{\varphi}$, $\dot{\delta}$ as states, only.

As mentioned in Section 3.3, the benchmark model is a well-known, practically validated and the mostly used bicycle model (e.g. [MPRS07, BMCP07, KSM08]). It is a LPV system with the constant forward velocity v as parameter given by

$$\boldsymbol{\pi} = \begin{pmatrix} \varphi \\ \delta \end{pmatrix}, \quad \boldsymbol{T}\,\ddot{\boldsymbol{\pi}} = -v\,\boldsymbol{K}_1\,\dot{\boldsymbol{\pi}} - \left[v^2\,\boldsymbol{K}_2 + g\,\boldsymbol{K}_0\right]\boldsymbol{\pi} + \boldsymbol{f}_{in}. \tag{4.62}$$

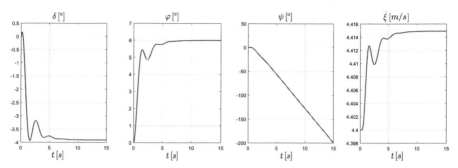

Figure 4.9: Simulation results for Scenario 7: Control of φ by only u_φ

g is the gravity constant, $\boldsymbol{T}$ is the mass matrix and $\boldsymbol{K}_1$, $\boldsymbol{K}_2$ and $\boldsymbol{K}_0$ are constant matrices. The linearised nonholonomic constraints are given by

$$\dot{\psi} = \frac{v\,\delta + \Delta\,\dot{\delta}}{L}\cos\lambda. \tag{4.63}$$

The behaviour of the system is characterised by an eigenvalue analysis. The eigenvalues of (4.62) are equal to the eigenvalues of the matrix

$$\boldsymbol{E} = \begin{pmatrix} \boldsymbol{0} & \boldsymbol{I} \\ -\boldsymbol{T}^{-1}\left(v^2\,\boldsymbol{K}_2 + g\,\boldsymbol{K}_0\right) & -v\,\boldsymbol{T}^{-1}\boldsymbol{K}_1 \end{pmatrix}, \tag{4.64}$$

which are discussed in Section 3.3 and illustrated in Figure 3.4.

For a better comparison, a constant forward velocity in the proposed model is considered. For comparison in the simulation, one can set $\ddot{r}^1 = 0$ and thus $\dot{r}^1 = const. = v$. However, for comparing the equations and linearisation, the terms from (4.20) need to be modified. The components of the mass tensor corresponding to $\ddot{\xi}$ are set to zero ($M_{D,1\mu} = M_{D,\mu1} = 0$). Further, all of the components of Λ corresponding to the first row of (4.20) are set to zero and therefore

$$M_{D,\alpha\mu} = \begin{pmatrix} M_{D,11} & 0 & 0 \\ 0 & & \\ 0 & & M_{D,\alpha'\beta'} \end{pmatrix}, \quad \Lambda_{\alpha\beta1} = 0. \tag{4.65}$$

This represents the motion of the system for a constant forward velocity $\dot{r}^1 = v$. Sine the objective is a LPV system however, the components of Λ corresponding to bilinear terms in $\dot{r}^2$ and $\dot{r}^3$ need to be set to zero resulting in

$$\Lambda_{\alpha\beta\mu'}(r) = \begin{pmatrix} \Lambda_{11\mu'} & \Lambda_{12\mu'} & \Lambda_{13\mu'} \\ \Lambda_{21\mu'} & & \\ \Lambda_{31\mu'} & & \boldsymbol{0} \end{pmatrix}. \tag{4.66}$$

Now the last two rows of (4.20) are decoupled from the first row and the equations of motion can be reduced to a system with only two coordinates r^2, r^3 as

$$M_{D,\alpha'\mu'}\,\ddot{r}^{\alpha'} = \Lambda_{11\mu'}\,v^2 + v\left(\Lambda_{1\beta'\mu'} + \Lambda_{\beta'1\mu'}\right)\dot{r}^{\beta'} - S^i_{\mu'}\,dU_i + u_{\mu'}. \qquad (4.67)$$

Using the Taylor approximation, (4.67) can be rewritten in the same form as (4.20) by

$$\bar{M}_{D,\alpha'\mu'}\,\ddot{r}^{\alpha'} + v\,K_{1,\beta'\mu'}\,\dot{r}^{\beta'} + \left(v^2\,K_{2,\alpha'\mu'} + g\,K_{0,\alpha'\mu'}\right)r^{\alpha'} = u_{\mu'} \qquad (4.68)$$

with

$$\bar{M}_{D,\alpha'\mu'} = M_{D,\alpha'\mu'}\big|_{r^\alpha=0}\,, \quad K_{1,\beta'\mu'} = -\left.\left(\Lambda_{1\beta'\mu'} + \Lambda_{\beta'1\mu'}\right)\right|_{r^\alpha=0} \qquad (4.69)$$

$$K_{2,\alpha'\mu'} = -\left.\frac{\partial \Lambda_{11\mu'}}{\partial r^{\alpha'}}\right|_{r^\alpha=0}\,, \quad K_{0,\alpha'\mu'} = \frac{1}{g}\left.\frac{\partial S^i_{\mu'}\,dU_i}{\partial r^{\alpha'}}\right|_{r^\alpha=0} \qquad (4.70)$$

The eigenvalues of the resulting system given by (4.67) are calculated by (4.64) and can be plotted with respect to the forward velocity v, similar to the picture in Figure 3.4. Figure 4.10 illustrates the eigenvalues of the reduced system using the parameters form Table 4.2. Note that the shape of the curves are similar, yet not identical to the curves from Figure 3.4. The difference is mainly due to the assumptions in the physics of the considered vehicle model, for instance the approximation (4.45) and neglecting the pitch. Also, the proposed model in Section 4.4 is a two-body system with a rotating front wheel while the benchmark model considers a 4-body model. Yet, the expected eigenmodi as well as desired behaviour of the eigenvalues are present.

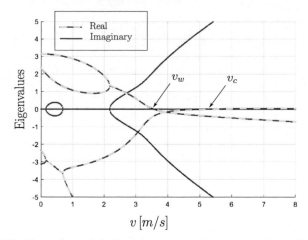

Figure 4.10: Eigenvalues of the linearised system for constant forward velocity v

4.6 Model simplification and extension

Model simplification

The equations of motion given by (4.20) for complex systems, such as a bicycle, are very large and contain many nonlinear terms. Therefore, for some applications such as online optimisation where the simulation time is critical, a simplified models is necessary. The EoMs given by (4.20) can be simplified to be described only by constant tensors. To this end, the components of $\Lambda_{\mathcal{D}}$ are linearised using the Taylor series of order one while in the components of $M_{\mathcal{D}}$ the variables r are substituted by zero. This yields the simplified equations of motion as

$$M_{\mathcal{D},\alpha\mu}(r^\alpha = 0)\,\ddot{r}^\alpha = \bar{\Lambda}_{\mathcal{D},\alpha\beta\mu}\,\dot{r}^\alpha\,\dot{r}^\beta - g\,G_\mu + u_\mu \tag{4.71}$$

where similar to (4.70),

$$G_\mu = \frac{1}{g}r^\gamma\,\frac{\partial\,S^i_\mu\,dU_i}{\partial r^\gamma}\bigg|_{r^\alpha=0} \quad,\quad \bar{\Lambda}_{\mathcal{D},\alpha\beta\mu} = \Lambda_{\mathcal{D},\alpha\beta\mu}(r^\alpha = 0) + r^\gamma\,\frac{\partial\Lambda_{\mathcal{D},\alpha\beta\mu}}{\partial r^\gamma}\bigg|_{r^\alpha=0}. \tag{4.72}$$

The major advantage of (4.71) is the fact that only constant tensors are used to describe the motion of the system. Once initialised with the vehicle parameters, a simulation can be run much faster. This can be used for tasks where an online simulation is required such as online trajectory planning or model predictive control methods.

Equation (4.71) however, does not preserve the port-Hamiltonian structure of the system. For a structure preserving simplification, one may simplify the overall metric tensor using a second order Taylor approximation. The metric tensor $M_{\mathcal{D}}$ can be approximated by

$$M_{\mathcal{D},\alpha\beta} \simeq \tilde{M}_{\mathcal{D},\alpha\beta} := \bar{M}_{rk0,\alpha\beta} + \bar{M}_{\mathcal{D}1,\alpha\beta\gamma}\,r^\gamma + \frac{1}{2}\,\bar{M}_{\mathcal{D}2,\alpha\beta\gamma\nu}r^\gamma r^\nu \tag{4.73}$$

leading to the equations

$$\tilde{M}_{\mathcal{D},\alpha\mu}\,\ddot{r}^\alpha = \tilde{\Lambda}_{\mathcal{D},\alpha\beta\mu}\,\dot{r}^\alpha\,\dot{r}^\beta - g\,G_\mu + \tau_\mu \tag{4.74}$$

where $\tilde{\Lambda}_{\mathcal{D},\alpha\beta\mu}$ is determined using $\tilde{M}_{\mathcal{D},\alpha\mu}$ as from (4.21). To verify and compare these equations, a simulation is run where the vehicle starts with a forward velocity of $v = 4.17\,\mathrm{m/s}$ and a torque disturbance is applied to the scenario 4 from the previous subsection. The velocity was intentionally chosen to illustrate the difference. The results illustrated in Figure 4.11 show that the difference is marginal and the general behaviour of the system is almost unchanged. Note that the difference becomes more distinct, yet acceptable for lower velocities.

Model extension

As mentioned before, one of the advantages of the proposed model is the extensibility by additional parts, for instance a leaning rider or further actuators for active stabilisation at low forward velocities.

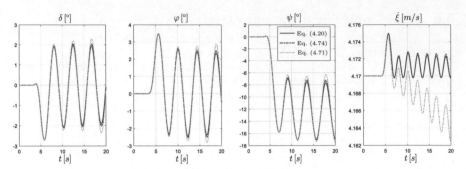

Figure 4.11: Comparison of the simplified models with the original model

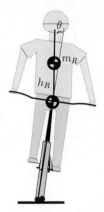

Figure 4.12: Schematics of the bicycle with the rider represented by an inverted pendulum

For demonstration, the introduced model is extended by an actuated inverted pendulum representing a leaning rider. It is represented by a mass point m_R attached to a massless link with the length h_R as illustrated in Figure 4.12. The leaning torque is notated by u_θ. The leaning angel relative to the rear frame is θ. For determining the kinetic energy, the metric tensor and the potential energy, the position vector of m_R is required, that is

$$_Z\boldsymbol{w}_R = {}_Z\boldsymbol{w}_c + {}^{ZR}\boldsymbol{R} \begin{pmatrix} 0 \\ 0 \\ h_R \end{pmatrix} , \quad _B\boldsymbol{\Omega}_R = {}_B\boldsymbol{\Omega}_c + \begin{pmatrix} \dot{\theta} \\ 0 \\ 0 \end{pmatrix} . \tag{4.75}$$

Calculating the metric tensor and the potential energy, the overall equations of motion are given by (4.20). The system parameters used for the simulation are taken from Table. 4.2, except for $m = 63\,\mathrm{kg}$, $m_R = 20\,\mathrm{kg}$ and $h_R = 0.2\,\mathrm{m}$. Note that m_R is not directly the mass of the rider but the mass of the upper body or the *moving part* of the body. Simulations are run with a state feedback generating the leaning torque of the rider u_θ. The initial velocity is set to $v = 4.4\,\mathrm{m/s}$, which is within the self-stability range. The bicycle is subject to a side-kick as in scenario 4 from the previous subsection. Two different sets of parameters are applied to obtain the state feedback given by

$$
\begin{aligned}
u_{\theta,1} &= -50\,\theta - \dot\theta - 2\,\delta - 4\,\dot\delta - 10\,\varphi - \dot\varphi \\
u_{\theta,2} &= -150\,\theta - 150\,\dot\theta - \delta - 10\,\dot\delta - 10\,\varphi - \dot\varphi.
\end{aligned}
\tag{4.76}
$$

The simulation results with the first feedback $u_{\theta,1}$ are illustrated in Figure 4.13. The applied torque for the leaning of the pendulum is able to compensate the disturbance torque and bring all of the angles back to zero. More interestingly, if $u_{\theta,2}$ is applied, the controller is not able to maintain zero angles any more. Yet, the system converges to an equilibrium that is a circular motion. The results of this simulation are shown in Figure 4.14. Note that no controller were designed to achieve any closed loop behaviour. The objective of these simulations is mainly to demonstrate the ability of the model to reproduce a realistic physical behaviour of the vehicle.

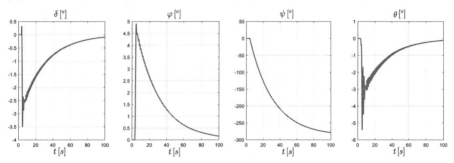

Figure 4.13: Simulation of the bicycle with rider and $u_{\theta,1}$ as in (4.76)

The effects of idealisation

The model introduced in this chapter is intended to be used for trajectory planing and tracking control and neglects some physical facts to keep the equations as simple enough. An important issue that has a considerable effect on the system dynamics is friction. Assuming viscose friction in the steering joint and roll friction of the tires extends the matrix $\boldsymbol{R}$ in (2.86) as $\boldsymbol{R} - \boldsymbol{F}$. $\boldsymbol{F} = \boldsymbol{F}^T$ is a positive definite matrix that contains the friction constants. A dry friction, however, is much more challenging to include into the system and is beyond the scope of this work. There are already investigations on dry friction in port-Hamiltonian systems such as [LLG05]. Neglecting the dry friction may indeed have an remarkable effect on the vehicle dynamics.

Another interesting entity is the lateral slip of the wheels. This would change the kinematics of the vehicle and therefore the nonholonomic constraints (4.61). To include lateral slip

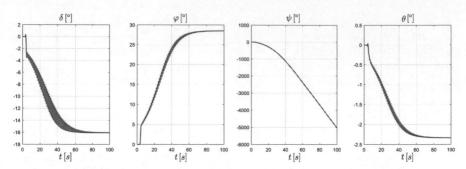

Figure 4.14: Simulation of the bicycle with rider and $u_{\theta,2}$ as in (4.76)

of the wheels, existing approaches of tire modelling from vehicle control literature (e.g. [CP10]) can be used. In fact, the lateral slip can be included by relaxing the nonholonomic constraints to determine a new A. As stated in [VB10], a more realistic tire model leads to a change in the self-stability behaviour of the bicycle. Adding friction even leads to instability of the uncontrolled bicycle in the velocity range, where it is supposed to be self-stable. Following on the existing literature in modelling an control of bicycles, and, since the application of the model in this work is the autonomous driving, these effect are neglected in this thesis, too.

5 Motion planning for autonomous two-wheeled vehicles

One of the use cases for the model introduced in Chapter 4 is model-based motion planning for autonomous driving. In the literature, motion planning is often divided into two or more hierarchical layers for each of which usually a different approach is considered. In general, the higher the layer in the hierarchy, the finer the decision to be made in that layer. It is usual that a strategical or tactical planning at the highest level makes abstract decisions such as *keep lane* or *turn right*. As a middle layer, a path planning layer is involved that is required to obtain the (in some sense) best collision-free path between two points. In this context, the lowest layer is often considered to be the trajectory planning that is supposed to create trajectories with regard to a vehicle model to achieve a pre-defined objective. Although tasks such as path planning lead to very interesting methodic and algorithmic challenges, this chapter focuses only on trajectory planning that is incorporating the vehicle model.

In Section 5.1, a general introduction to trajectory planning for autonomous vehicles is given, followed by a brief explanation of the applied planning approach. That is, in Section 5.2, used to create model-based optimal trajectories for an autonomous TWV. Thereby, several vehicle models regarding model complexity and completeness as described in Chapter 4 are used. The main objective is a demonstration of usability of the vehicle model developed in Chapter 4 for trajectory planning as well as an investigation of different vehicle models with respect to optimal trajectory planning. To this end, a scenario is devised in which the bicycle is supposed to reach a specific point on the ground avoiding two obstacles. A simulation setup is used to verify the resulting trajectories by applying them to a two-degree-of-freedom control loop and the main differences are emphasized.

5.1 Introduction to trajectory planning for autonomous vehicles

Trajectory planning for mobile vehicles has been a matter of research for a long time. Preliminaries on motion planning for mobile robots can be found e.g. in [Lau98]. A good survey on the recent developments in motion planning for autonomous vehicles is given in [PCY+16]. [HPKK14] gives an overview on some key methods for model-based motion planning used for mobile robots. Research on trajectory planning for bicycles specifically has not been as extensive as those for mobile robots or other vehicles, as also stated in [YCSH14].

Yet, many aspects are similar and can be reasonably transformed. [HSF08] discusses general achievable trajectories for motorcycles and [SHB12] uses a rigid body model for creating trajectories using dynamic inversions. [YCSH14] uses the nonlinear model from [GM95] to design optimal trajectories.

In general, the optimal control problem

$$\min_{\boldsymbol{u}(t), \boldsymbol{x}(t)} \tilde{\mathcal{J}}_b(\boldsymbol{u}_0, \boldsymbol{x}_0, \boldsymbol{u}_F, \boldsymbol{x}_F, t_0) + \int_{t_0}^{t_F} \tilde{\mathcal{J}}_I(\boldsymbol{u}(t), \boldsymbol{x}(t), t) \, dt$$

$$\text{s. t.} \quad \dot{\boldsymbol{x}} = \boldsymbol{f}(\boldsymbol{x}, \boldsymbol{u}, t) \,, \, \{\boldsymbol{x}(t), \, \boldsymbol{u}(t)\} \in \mathcal{C} \subseteq \mathcal{X} \times \mathcal{U}$$

$$\boldsymbol{x}_0 = \boldsymbol{x}(t = t_0) \,, \, \boldsymbol{x}_F = \boldsymbol{x}(t = t_F) \tag{5.1}$$

has to be solved to determine optimal trajectories where $\tilde{\mathcal{J}}_b$ is the objective functional at the boundaries and $\tilde{\mathcal{J}}_I$ is the so-called *running cost*. t_0 and t_F are the starting time and the final time correspondingly. $\boldsymbol{f}(\boldsymbol{x}, \boldsymbol{u}, t)$ describes the general dynamics of the considered system and $\mathcal{C}$ is the generalized manifold on which the constraints are satisfied. $\mathcal{X}$ and $\mathcal{U}$ correspond to the constraints on the states and inputs, $\boldsymbol{x}$ and $\boldsymbol{u}$, accordingly.

In the literature, for instance in [DBDW05] or [Kel17], three different types of algorithms for solving this problem are identified:

1. Dynamic Programming uses the principle optimality of curves to determine the optimal trajectory that leads to a Jacobi-Hamilton-Bellman Partial Differential Equation (PDE) which needs to be solved.

2. Indirect methods use necessary optimality conditions to derive a boundary value problem as a set of ODEs, which are solved by numeric integration methods.

3. Direct methods transform the infinite-dimensional problem into a finite-dimensional Nonlinear Program (NLP)

$$\min_{\boldsymbol{z}} \mathcal{J}(\boldsymbol{z})$$

$$\text{s. t.} \quad \boldsymbol{b}_l \leq \begin{pmatrix} \boldsymbol{z} \\ \boldsymbol{C}(\boldsymbol{z}) \\ \boldsymbol{A}\boldsymbol{z} \end{pmatrix} \leq \boldsymbol{b}_u \tag{5.2}$$

that is solved using numerical methods. $\boldsymbol{z}$ is referred to as the vector of the decision variables or the optimisation variables and $\boldsymbol{C}(\boldsymbol{z})$ summarises the nonlinear constraints. $\boldsymbol{b}_l$ and $\boldsymbol{b}_u$ are the lower and upper bounds for the optimisation variables respectively. Note that $\boldsymbol{C}(\boldsymbol{z})$ may also contain equality constraints in which case the corresponding entries of $\boldsymbol{b}_l$ and $\boldsymbol{b}_u$ are equal.

Same categorization is found in [PCY+16] and other related literature. A survey on different numerical methods used for trajectory optimization can be found in [Bet98].

There is already a variety of highly efficient numerical algorithms for solving the NLP (5.2), for instance `fmincon` from MathWorks® or `IPOPT`[1] which makes the third type of algorithms very attractive.

Transcription: The act of transforming the optimal control problem into the NLP is referred to as discretization or more often as *transcription*. [Kel17] distinguishes between two transcription methods: *Shooting* and *Simultaneous*. First one uses an approximation of the input using predefined functions and numeric integration of the system dynamics to calculate the corresponding states. The latter, however, approximates both the input and the state of the systems and uses equality constraints to satisfy the system dynamics at certain (collocation) points. The method used in Section 5.2, *Direct Collocation*, is an special case of the second method, where the input is approximated as a piecewise linear function and the states as piecewise quadratic. More details on different transcription methods and their application to trajectory optimization can be found in [Kel17]. Furthermore, one may distinguish between the methods for numerical integration of the system dynamics. The trajectory planning discussed in this chapter uses the *trapezoidal* integration that is defined by

$$x_1 = \int_{t_0}^{t_1} f \, dt \;\Rightarrow\; x_1 \simeq x_0 + \frac{1}{2} T_s \left(f_1 + f_0 \right) \tag{5.3}$$

with the sample time T_s. The objective functional is also approximated using the sum

$$\min_{u(t),x(t)} \int_{t_0}^{t_1} \tilde{\mathcal{J}} \, dt \simeq \min_{u_0,u_1,\cdots x_0,x_1,\cdots} \sum_{k=0}^{N-1} \frac{1}{2} T_s \left(\tilde{\mathcal{J}}_k + \tilde{\mathcal{J}}_{k+1} \right). \tag{5.4}$$

The decision variables are taken to be the inputs and the states at all time instances, for instance

$$z = \begin{bmatrix} z_0 \\ z_1 \\ \vdots \\ z_{N-1} \end{bmatrix} := \begin{bmatrix} u_0 \\ x_0 \\ u_1 \\ x_1 \\ \vdots \\ u_{N-1} \\ x_{N-1} \end{bmatrix} \tag{5.5}$$

with N the number of the collocation points.

The constraints are enforced for each node point k with $t_k = k\,T_s$ as

$$b_l \leq \mathbf{C}(x_k, u_k) \leq b_u \tag{5.6}$$
$$x_l \leq x_k \leq x_u \tag{5.7}$$
$$u_l \leq u_k \leq u_u. \tag{5.8}$$

---

[1] `www.coin-or.org/Ipopt`

Note that in this case constant bounds and constraints are assumed, however variable constraints may also be applied which are different at each collocation point.

Interpolation: Once the optimisation is terminated successfully, the optimal values for z at the collocation points are known. To obtain a continuous-time trajectory, an interpolation is applied. In the method used in this work, direct collocation, the inputs $u(t)$ is obtained by a linear interpolation between the collocation points. The states $x(t)$ are determined using quadratic splines between the collocation points as

$$u(t) \approx u_k + \frac{t - t_k}{h_k}(u_{k+1} - u_k)\,, \quad t \in [t_k, t_{k+1}] \tag{5.9}$$

$$x(t) \approx x_k + f_k(t - t_k) + \frac{(t - t_k)^2}{2\,T_s}(f_{k+1} - f_k)\,, \quad t \in [t_k, t_{k+1}]\,. \tag{5.10}$$

The nonlinear solver `fmincon` included in Matlab® is applied to create optimal trajectories as will be explained in the next section. Note that the trajectory optimisation problem has to be re-formulated to match the problem form required by the solver for which the *Trajectory-Optimization Toolbox* developed by Dr. Matthew Kelly[2] is used that is described in [Kel17].

5.2 Trajectory planning for a two-wheeled vehicle by direct collocation

The considered scenario is an S-curve manoeuvre in presence of obstacles which are to be avoided. A path X is a set of points on the x-y-plane. The task of the trajectory planning is to find a trajectory with the corresponding path to navigate the vehicle safely around the obstacles and, then, reach the desired goal point $(70\,\mathrm{m}, 70\,\mathrm{m})$ with the same orientation angle.

Since we use a collocation method and consider the ground track of the vehicle too, the decision variables are given by $z = [x^T, X^T, u^T]^T$ with

$$x = \begin{bmatrix} q \\ \dot{r} \end{bmatrix} = \begin{bmatrix} \xi \\ \varphi \\ \delta \\ \psi \\ \eta \\ \dot{\xi} \\ \dot{\varphi} \\ \dot{\delta} \end{bmatrix} \text{ and } u = \begin{bmatrix} u_\xi \\ u_\varphi \\ u_\delta \end{bmatrix}\,.$$

The torque u_φ which is directly applied to the leaning angle is taken to be zero in the following to consider the under-actuated vehicle model as will be explained in Chapter 6.

---

[2]www.matthewpeterkelly.com

Note that the path variables X are explicitly considered in the model and are, therefore, included in the optimisation problem by

$$\dot{x} = \dot{\xi}\cos\psi\,, \quad \dot{y} - \dot{\xi}\sin\psi. \tag{5.11}$$

Four different vehicle models are considered and the trajectory planing results are compared with regard to model complexity and physical extensiveness as explained in Chapter 3. The *Trailed* model is the most detailed and, thus, the most complex model which includes the geometric entities leading to the self-stability such as Δ and d_s. The name *Trailed* is chosen to emphasise the fact that the trail is not neglected, contrary to the *Naive* model, which neglects the trail and some other physical effects, as explained in Chapter 4. Both models are given by (4.20), in the *Naive* model however, $\Delta = d_s = \epsilon = 0$ is set. Furthermore, a simplified model is considered created by simplification of the *Trailed* model as from (4.74). Finally, a further simplified model from the *Trailed* model is considered that is described by constant tensors as from (4.71).

As mentioned in the previous section, the direct collocation method uses a linear approximation of the inputs and quadratic approximation of the states. This necessarily leads to a deviation from the physical behaviour for complex systems. For a naturally unstable system such as the bicycle, this may lead to a large deviation of the states, once the resulting optimal input is applied in an open loop. For this reason, the structure illustrated in Figure 5.1 is used for verification and comparison of the trajectories. The optimisation problem is created using each vehicle model and solved. Then, the optimal input $u^\star$ is applied to the full bicycle model (*Trailed*) as a feed-forward in addition to a linear feedback using the deviation from the optimal states $x - x^\star$. Note that this feedback is a simple linear controller and not designed to track the desired state, rather just to prevent the bicycle from falling over during the manoeuvrer. In the closed loop simulations described below, the state feedback

$$u_\delta = -1200\varphi - 800\delta - 1600\dot{\varphi} - 800\dot{\delta}. \tag{5.12}$$

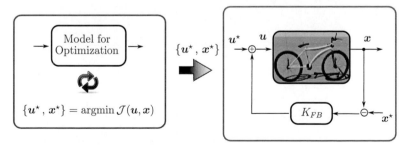

Figure 5.1: Schematics of the trajectory optimisation for the bicycle

As illustrated in Figure 5.2, a predefined nominal path $\boldsymbol{X}_{\text{Nom}}$ is assumed to be given as distributed points between the starting point and the goal point as

$$\boldsymbol{X}_{\text{Nom}} = \{(x,y) \mid x \in [0,70], \, y \in [0,70]\} \tag{5.13}$$

that is a straight line, turning 90° to the left, turning 90° to the right and, then, ending in the goal point. The number of collocation points is taken to be $N = 10$. The initial states are

$$\dot{\xi}(0) = 4.3\,\text{m/s}\,, \; \varphi(0) = \delta(0) = \psi(0) = 0\,, \; \boldsymbol{X}(0) = (0,0) \tag{5.14}$$

and the desired terminal values are

$$\dot{\xi}_F = 4.3\,\text{m/s}\,, \; \varphi_F = \delta_F = \psi_F = 0\,, \; \boldsymbol{X}_F = (70,70). \tag{5.15}$$

It is supposed that two obstacles are located at $\boldsymbol{X}_{\text{O},1} = (8,15)$ and $\boldsymbol{X}_{\text{O},2} = (70,40)$. The transcribed system dynamics from the corresponding vehicle model is taken as the nonlinear constraint, where for each model, the corresponding tensors from Chapter 4 are used. The objective functional is defined by

$$\mathcal{J} = \mathcal{J}_{\text{P}} + \mathcal{J}_{\text{u}} + \mathcal{J}_{\text{c}} + \mathcal{J}_{\text{d}} + \mathcal{J}_{\text{O}}\,, \tag{5.16}$$

$$\mathcal{J}_{\text{P}} = K_p \sum_{n=1}^{N-1} \|\boldsymbol{X}(n) - \boldsymbol{X}_{Nom}(n)\|_2 +$$

$$K_{p,F} \,\|\boldsymbol{X}(N) - \boldsymbol{X}_{\text{Nom}}(N)\|_2 + K_{\psi,F} \,(\psi(N) - \psi_F)^2\,,$$

$$\mathcal{J}_{\text{u}} = K_{\text{u}} \sum_{n=1}^{N-1} \|\boldsymbol{u}(n)\|_2\,,$$

$$\mathcal{J}_{\text{c}} = K_{c,\varphi} \sum_{n=1}^{N} \varphi(n)^2 + K_{c,\delta} \sum_{n=1}^{N} \delta(n)^2\,,$$

$$\mathcal{J}_{\text{d}} = K_{d,\varphi} \sum_{n=1}^{N} \dot{\varphi}(n)^2 + K_{d,\delta} \sum_{n=1}^{N} \dot{\delta}(n)^2\,,$$

$$\mathcal{J}_{\text{O}} = K_{\text{O}} \sum_{n=1}^{N} \frac{1}{\|\boldsymbol{X}(n) - \boldsymbol{X}_{\text{O},1}\|_2 + \|\boldsymbol{X}(n) - \boldsymbol{X}_{\text{O},2}\|_2 + 10^{-12}}\,. \tag{5.17}$$

with $\| \cdot \|_2$ the matrix 2-norm. The objective functional contains different terms. The first one, $\mathcal{J}_{\text{p}}$, penalises the distance to the nominal path weighted by $K_p > 0$. It also includes the path terminal costs weighted by $K_{p,F} > 0$ as well as the terminal cost on the vehicle orientation weighted by $K_{\psi,F} > 0$. This term makes sure that the vehicle orientation is as desired at the end of the manoeuvrer. This is even more important once trajectory planning is supposed to run in a cyclic manner due to the nonholonomic constraints on the vehicle kinematics. A wrong vehicle orientation at the beginning of the next cycle may lead to undesired trajectories or, in worst case, infeasibility of the follow-up problem. $\mathcal{J}_{\text{u}}$ penalises the input effort. $\mathcal{J}_c$ penalises the leaning angle and the steering angle to prevent the bicycle from falling over and $\mathcal{J}_d$ is a measure for riding comfort and prevents sudden change of the leaning or steering angles in order to provide smoother functions for those angels. $\mathcal{J}_O$ makes sure that the distance of every collocation point to each obstacle is maximised. Note that the additive term 10^{-12} in the denominator prevents numerical instability. Note that the obstacles are only considered in the objective functional and not as hard constraints.

The lower and upper bounds for the variables z_n are taken to be (all SI-Units)

$$z_l = \begin{bmatrix} x_l, X_l, u_l \end{bmatrix}^T , \quad z_u = \begin{bmatrix} x_u, X_u, u_u \end{bmatrix}^T ,$$

$$x_l = \begin{bmatrix} 0 & -\dfrac{\pi}{3} & -\dfrac{\pi}{2.1} & -\pi & -10^{-16} & 3.7 & -20 & -20 \end{bmatrix}^T ,$$

$$x_u = \begin{bmatrix} \infty & \dfrac{\pi}{3} & \dfrac{\pi}{2.1} & \pi & 10^{-16} & 4.5 & 20 & 20 \end{bmatrix}^T ,$$

$$X_l = \begin{bmatrix} 0 & -1 \end{bmatrix}^T , \quad X_u = \begin{bmatrix} 75 & 76 \end{bmatrix}^T ,$$

$$u_l = \begin{bmatrix} -2 & 0 & -0.2 \end{bmatrix}^T , \quad u_u = \begin{bmatrix} 2 & 0 & 0.2 \end{bmatrix}^T . \tag{5.18}$$

Note that the constraints on η are taken to be almost zero (10^{-16}) to make sure that the initial guess strictly satisfies all constraints and the solver does not throw warnings or errors. Also, the middle term in u_l and u_u that corresponds to u_φ is taken to be constantly zero. This is due to the under-actuated nature of the bicycle. There are already approaches to add a direct torque to the leaning angle using different actuators such as CMG (e.g. [KYKY11], [GMÁF18], [WYLZ17]), in this chapter however, a conventional bicycle is considered where $u_\varphi = 0$. The weighting parameter are chosen to be

$$K_u = 10 , \quad K_P = K_{p,F} = 8 , \quad K_P = K_{\psi,F} = \frac{8 \cdot 180}{\pi} ,$$

$$K_{d,\varphi} = K_{d,\delta} = 20 , \quad K_{c,\varphi} = K_{c,\delta} = 1 . \tag{5.19}$$

Since the path coordinates are included in the decision variables the resulting optimisation problem is necessarily non-convex. Thus, a good initial guess is crucial for the feasibility and the quality of the solution. In the current scenario, the initial guess is created only by simulating the vehicle model and a left-and-then-right turn by applying impulse-shaped torques on the steering angel u_δ.

The resulting trajectories are optimal and satisfy the requirements at the collocation points, yet not along the entire vehicle motion. The results are illustrated in Figure 5.2. One can see that the optimisation results are very similar for all four models at the collocation points (marked by ○). The optimisation time is, however very different. The time required to solve the problem with the *Trailed* model is much higher than the one with the *Naive* and the shortest time was required for the problem with the simplified model. The average required times, obtained by running the same program at least 20 times under the same conditions, are listed in Table 5.1. Note that the absolute values are not of interest since those depend on the computer hardware and software the optimisation is run on. The comparison is still however legitimate since all problems were run under the same condition and on the same computer.

Model used for optimisation	Required average time in [s]
Trailed wit h(4.20)	165
Naive with (4.20)	66
Simplified with (4.74)	68
Simplified with (4.71)	42

Table 5.1: Comparing the average required time for the trajectory optimisation using different vehicle models

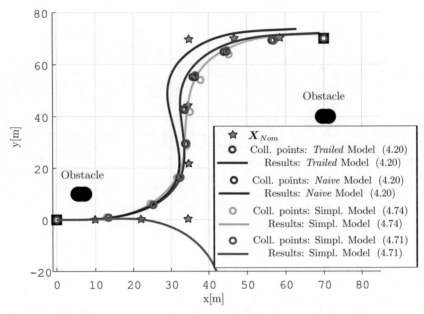

Figure 5.2: Comparing trajectory optimisation results by different vehicle models on the x-y-plane

Once the trajectories are applied to the *Trailed* bicycle model for simulation, it becomes obvious that the drawback of the low optimisation time is the error in the input and the states. As shown in Figure 5.2, the trajectories created by the *Trailed* model has the smallest deviation from the original path. Trajectories created by the simplified model according to (4.71) were not able to be stabilised by the simple state feedback (5.12). A similar pattern can be found out in the experimental investigations described in Chapter 8. The trajectories created by the simplified models are still usable, yet not accurate enough. In Figures 5.3, 5.4, 5.5 and 5.6 relevant states from each trajectory and the corresponding simulation results are illustrated. Red curves show the interpolated trajectories as the last step of the method direct collocation as explained above.

Note that the leaning angle φ cannot be tracked well since the feedback (5.12) is not designed to do so. In chapter 6, a nonlinear passivity-based trajectory tracking controller is introduced to tackle this issue. In summary, the comparison shows that different vehicle models can be used to create optimal trajectories for autonomous TWVs. The results differ with regard to the quality of the created trajectories, as well as the required computation time. Whether and how well a trajectory created by a specific vehicle model can be tracked by a simple state feedback applied to the most realistic model, is taken as a measure for the quality of that trajectory. Using a more complex model leads to a higher computation time and a better trajectory in this sense.

The results show that the *Trailed* model developed in Chapter 4 leads to the best trajectory with a rather high computation time, that is an expected fact. More interestingly, using the simplified vehicle model obtained by a quadratic approximation of the tensors involved in the *Trailed* model as given by (4.74), leads to trajectories with a similar quality to those created by the *Trailed* model. However, the computation time is similar to the case using the *Naive* model. Since the simplified model from (4.74), contrary to the *Naive* model, includes important physical effect of TWVs as explained in Chapter 4, it is reasonable to suggest, in conclusion, that this is the best practical choice for creating optimal trajectories for autonomous TWVs.

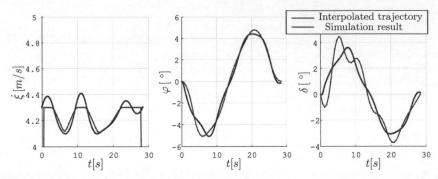

Figure 5.3: Relevant states of the closed loop simulation using optimal trajectories created by the *Trailed* model

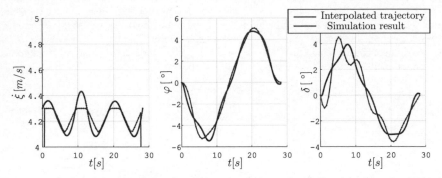

Figure 5.4: Relevant states of the closed loop simulation using optimal trajectories created by the *Naive* model

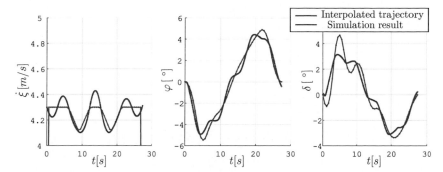

Figure 5.5: Relevant states of the closed loop simulation using optimal trajectories created by the simplified model with (4.74)

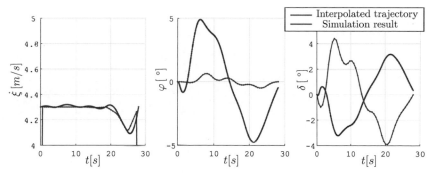

Figure 5.6: Relevant states of the closed loop simulation using optimal trajectories created by the simplified model with (4.71)

6 Controller synthesis for autonomous two-wheeled vehicles

Controller synthesis for autonomous TWVs is a challenging task, mainly due to the complex model and the dynamical characteristics, which has been paid attention to in the last years. Comparing to four-wheeled vehicles, there exist much fewer literature on the controller design for autonomous TWVs. One reason is likely to be the missing practical necessity of autonomous two-wheeled vehicles in the past.

Using the linear benchmark model from [MPRS07] and different modifications to it, several approaches have been proposed for the control of an autonomous bicycle. [HTI16] uses the benchmark model to design and analyse controllers for bicycle stabilisation. [YCSH14] uses the benchmark model for trajectory planing and tracking control. [EHP15] adds a pendulum to the benchmark model representing a rider and designed a linear controller for autonomous riding. [GMÁF18] adds a flywheel for active stabilisation of the leaning angle and used a linear controller for stabilising the system.

The benchmark model is only valid for constant velocities and different modifications do not deliver a model valid for all, especially small, velocities and path curve radii. Therefore, nonlinear approaches have been also investigated. [GM95] for instance, introduces the so-called *Naive* bicycle using the approach of constrained Lagrangian and designs an nonlinear controller based on dynamic inversion. [Zha14] adds a modification to that model to include some missing physical features, such as balancability at zero velocity, and designs a nonlinear controller. [US06] designs a fuzzy controller for the *Naive* bicycle model. [RRPGCG$^+$17] introduces an energy-optimal controller based on linearisation of the *Naive* bicycle. [Keo08] and [YKV$^+$14] add a flywheel to the *Naive* bicycle model for active stabilisation of the leaning angle. [WYLZ17] adds a flywheel to the bicycle model from [Zha14] for active stabilisation and designs a nonlinear controller.

Controllers designed for the benchmark model are only valid for constant forward velocities and cannot properly handle the time-varying nonlinear problem of trajectory tracking. Nonlinear approaches are mostly complex and a majority of them focuses mainly on balance control of bicycles, or tackle the balance stabilisation and the trajectory tracking in a separate manner. The contribution of the present work in this context is a unifying approach to modelling, optimal trajectory planning and trajectory tracking based on the systematic methods of passivity-based control. In this chapter, the Port-Hamiltonian (PH) formulation of the TWV model developed in Chapter 4 is used to design a passivity-based trajectory tracking controller. The main objective is to obtain a controller for tracking realisable trajectories for the under-actuated vehicle model.

PH-systems are particularly interesting to control engineering because of their natural passivity. A majority of existing methods to design Passivity-Based Controllers (PBCs) is, therefore, applied to PH-systems. Once a model is given as a PH-system, existing systematic approaches of PBC can be applied that are known to be inherently robust against model uncertainties, as stated by [vdSJ14].

In Section 6.1, the general approach of PBC is described and the available literature is summarised. Thereby, two particular methods of PBC are explained, which are Iterconnection and Damping Assignment (IDA) and Generalised Canonical Transfomration (GCT). In Section 6.1.1, IDA-PBC is described briefly since it is a fundamental approach of PBC. Furthermore, the approach of IDA for handling under-actuated mechanical systems is the basis of the controller design for the autonomous TWV described in Section 6.3.3. In Section 6.1.2, GCT is explained in detail which is the approach modified later in this chapter for stabilising and trajectory tracking of the autonomous TWV. In Section 6.2, the general approach of designing a GCT-based controller for trajectory tracking is explained which is a time-varying problem. In the available literature, GCT is often used for trajectory tracking control because of the ability to handle time-varying systems. However, fully-actuated systems with no gravity are considered, only. IDA is, in contrast, widely used for stabilisation of under-actuated mechanical systems. In this chapter, a GCT-based trajectory tracking controller is proposed for mechanical systems with gravity, which is then, using the methods from IDA modified to apply to the under-actuated TWV model.

In Section 6.3 trajectory tracking controllers are derived for different models of two-wheeled vehicles and applied to achieve different objectives. Both the pseudo-Hamiltonian model of the nonholonomic system as well as the corresponding equations for the controller design become larger and more difficult with the growing model complexity. For a better understanding of the approach therefore, in Section 6.3.1, the controller is first derived for a simpler system, that is a planar two-wheeled vehicle or a so-called single-track model from Section 4.3. Then, in Section 6.3.2, while the general pattern of the design remains unchanged, specific equations are adjusted to re-do the controller design for the more complex system, that is the self-stable bicycle. Thereby, the bicycle model from Chapter 4 is considered which is, however, first assumed to be fully actuated. Although this assumption is not valid for regular bicycles or motorcycles, the systematics of the controller synthesis remains unchanged. Besides, fully actuated TWVs, namely with an additional actuator applying torque on the leaning angle, for instance a CMG, have been developed and are partly commercially available such as *Lit Motors*[1]. In the literature, there are also already approaches for modelling and control of such a vehicle as for instance by [KYKY11] and [YKV+14]. Different Scenarios are considered distinguished by the type of the desired trajectories and simulations show satisfying results. Furthermore, Monte-Carlo simulations are run to demonstrate the robustness of the closed loop against parameter uncertainties. In Section 6.3.3, the controller is modified to be applied to the more realistic, under-actuated

---

[1] www.litmotors.com

bicycle model that assumes no direct torque on the leaning angle. The same scenarios form Section 6.3.2 are run which deliver satisfying results. The controller introduced in Section 6.3.2 is generally designed for tracking the coordinates of the configuration space of a vehicle model. In Section 6.3.4, the controller is applied within an extended loop structure, that is furthermore able to track a ground path by modifying the desired values. Finally in Section 6.3.5, simulations validate the tracking control of optimal trajectories created as explained in Chapter 5.

6.1 Passivity-based control

The fundamental entity in the synthesis of a passivity-based controller is energy. A system is considered as a energy-converting unit which may consist of several sub-systems. Those themselves are again considered as energy-converting units which interact with each other exchanging energy. In this sense, a controller is interpreted as another sub-system that may influence the overall energy. The main idea of PBC is now to *shape* the energy of the closed-loop system using a proper controller to achieve desired dynamical behaviour [OVdSMM01]. To that end, a feedback law is computed to make sure the closed-loop is a passive system with a desired overall storage function, which is usually the overall energy in a physical system. The objective is thereby to choose the energy modification in the system such that a desired dynamical behaviour is achieved, for instance the stability of a particular equilibrium. Here, the passivity properties of PH-systems and their interconnection are often used. The energy-based nature of PBC leads to the closed-loop being robust against parameter variations according to [vdSJ14].

The general approach of PBC is explained in [OS89]. PBC is a control synthesis method which uses the characteristics of passive systems for stabilising a, in general, nonlinear system of the form (2.4). There exist a variety of systematic methods to solve such a stabilisation problem such as IDA (e.g. [OSGEB02]), Immersion and Invariance (I&I) (e.g. [AO03], [SAOM13]) and GCT (e.g. [FS01b]). IDA uses the interconnection of PH-systems to shape the energy of the closed loop while I&I uses invariant subspaces to maintain the passivity of the closed loop. Using GCT, a large class of dynamical systems can be tackled, including time-varying systems. In fact, [DS12a] claims that the IDA is a special case of GCT. GCT is the method used as the basis for the trajectory tracking controller design for autonomous TWVs in this chapter. More explanations on the concept of PBC can be found in [OVdSMM01] and [BIW91]. Further in this chapter, IDA and GCT are explained in detail. For more details on I&I, please refer e.g. to [AO03].

IDA-PBC was introduced in [OSGEB02]. [OGC04] explains this method more detailed and approaches under-actuated mechanical systems. In [VOB+07], the problem of solving necessary partial differential equations for the synthesis is approached and a controller is designed for a rotational pendulum. Further examples can be found in [AOAM05]. The main idea of IDA-PBC is to design a controller which forms the energy function of the closed loop system into a desired shape while the passivity of the closed loop system is guaranteed ([OVME02]).

The basic method of PBC is appropriate for the control of mechanical systems described by the Euler-Lagrange equations. Shaping only the potential energy, passivity-based control preserves the Euler-Lagrangian structure ([OVME02]). While shaping the total energy function of the closed loop system including the kinetic energy, preservation of the Euler-Lagrangian structure cannot be guaranteed any more. IDA-PBC can be interpreted as a generalised passivity-based control for PH-systems, which constitute a superset of systems described by Euler-Lagrange equations ([OVME02]). Some examples of IDA-PBC applications can be found in [GEORA01], [VOB⁺07] and [AOAM05]. IDA-PBC is often used to control under-actuated mechanical systems, e.g. in [OVME02], [MAOV06] and [GEvdS04]. In [KVL10], IDA-PBC is applied to design a controller for trajectory tracking.

The idea and the derivation of GCT is described in detail in [FS01b]. In principal, a structure preserving coordinate transformation is applied, yielding an new port-Hamiltonian system with a different (desired) Hamiltonian. In [FS99] and [FS01a], GCT is used to design controllers for systems with nonholonomic constraints. In [FS02] the controller is extended by an integral action for asymptotic stabilisation of electro-mechanical systems.

One of the advantages of GCT is the ability to handle time-varying system. Another noticeable advantage is that this method can be applied multiple times to achieve different objectives, in a way that a GCT-synthesis is applied to the resulting closed loop of the previous synthesis. This is explained more clearly in the following section. This makes this approach suitable for trajectory tracking control design as applied in [FSS04] and [FT08]. To design a trajectory tracking control using GCT, first, a time-varying error system is generated in a structure preserving manner that reduces the tracking problem to a stabilisation problem. Then, a second transformation stabilises the error system rendering the tracking error asymptotically zero.

Given a plant as a PH-system (2.21), the general synthesis approach of a PBC is to find a feedback law preserving the PH structure in the closed-loop with a different, desired Hamiltonian H_d, as well as a new structure matrix $\boldsymbol{J}_d$ and a new damping matrix $\boldsymbol{R}_d$. If this succeeds the closed loop is Lyapunov-stable at the minimum of the overall Hamiltonian. The process of changing the overall energy of the close-loop is called *Energy Shaping*. To achieve asymptotic stability of the desired equilibrium, a next step is required which is called *Damping Injection*. Here, the output of the passive closed-loop system is fed back negatively. Below, these concepts are explained further.

Clarification: Please note that for designing a PBC using PH-systems, usually the physical damping in the system, such as friction, is neglected in the modelling. The result is a passive system that is stable at the minimum of the Hamiltonian, however not necessarily asymptotically stable. Using *Energy Shaping*, the minimum of the Hamiltonian of the closed loop is shifted to a desired point. For asymptotic stability, in this case, an additional damping is then required that is considered in *Damping Injection*. The physical system, however, may be already asymptotically stable at the minimum of the new Hamiltonian due to the physical damping in the system. Therefore, *Damping Injection* is not a necessary, but a sufficient condition for asymptotic stability of the closed loop.

Energy Shaping
In the first part of a controller synthesis using PBC, the overall energy of the closed loop system is *shaped* into a desired from. Thereby, the natural passivity of PH-systems is used. Assuming a lossless PH-system (2.37), the objective is to find a control law $u(x,t)$ which makes the closed loop a passive system with a desired energy function implying desired dynamical behaviour. Many PBC approaches such as GCT and IDA make sure the closed loop is also a PH-system to maintain passivity. Defining the desired interconnection

$$J_d(q,p) = \begin{bmatrix} 0 & J_1(q) \\ -J_1^T(q) & J_2(q,p) \end{bmatrix},$$ (6.1)

the so-called *matching equation* is derived as

$$\begin{bmatrix} 0 & I \\ -I & 0 \end{bmatrix} \begin{bmatrix} \frac{\partial^T H(q,p)}{\partial q} \\ \frac{\partial^T H(q,p)}{\partial p} \end{bmatrix} + \begin{bmatrix} 0 \\ G_u(q) \end{bmatrix} u \stackrel{!}{=} \begin{bmatrix} 0 & J_1 \\ -J_1^T & J_2 \end{bmatrix} \begin{bmatrix} \frac{\partial^T H_d(q,p)}{\partial q} \\ \frac{\partial^T H_d(q,p)}{\partial p} \end{bmatrix}.$$ (6.2)

In general in PBC using PH-systems, the matrix Equation (6.2) results in two equations:

$$\frac{\partial^T H(q,p)}{\partial p} = J_1(q) \frac{\partial^T H_d(q,p)}{\partial p}$$ (6.3)

and

$$-\frac{\partial^T H(q,p)}{\partial q} + G_u(q) \cdot u = -J_1^T(q) \frac{\partial^T H_d(q,p)}{\partial q} + J_2(q) \frac{\partial^T H_d(q,p)}{\partial p}.$$ (6.4)

Assume H_d is chosen such that (6.3) is satisfied. For a fully actuated system, G_u^{-1} exists and therefore the equation above can be rewritten to yield u_{es} as

$$u_{\mathrm{es}} = G_u^{-1} \left(\frac{\partial^T H(q,p)}{\partial q} - J_1^T \cdot \frac{\partial^T H_d}{\partial q} + J_2 \cdot \frac{\partial^T H_d}{\partial p} \right).$$ (6.5)

For under-actuated systems, G_u has a rang defect and cannot be inverted. It is however possible to find an energy shaping control input using the pseudo-inverse as explained later in this chapter.

Damping Injection
Applying the controller from *energy shaping* results in a new passive system with a desired stable equilibrium. Therefore, to achieve asymptotic stability of the closed loop at the said equilibrium, the passive output needs to be fed back negatively. First, the passive output is determined as

$$y_{\mathrm{passive}} = G_{\mathrm{pH}}^T \cdot \begin{bmatrix} \frac{\partial^T H_d}{\partial q} \\ \frac{\partial^T H_d}{\partial p} \end{bmatrix}.$$ (6.6)

To adjust the *damping* in the system, the passive output is fed back using a matrix $\boldsymbol{K}_d = \boldsymbol{K}_d^T > \boldsymbol{0}$. Thus, the control input corresponding to damping injection is given as

$$\boldsymbol{u}_{\mathrm{di}} = -\boldsymbol{K}_d \cdot \boldsymbol{y}_{\mathrm{passive}} = -\boldsymbol{K}_d \cdot \boldsymbol{G}_u^T \frac{\partial^T H_d}{\partial \boldsymbol{p}}. \tag{6.7}$$

Figure 6.1 illustrates different parts of a PBC consisting of energy shaping and damping injection.

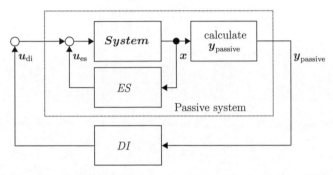

Figure 6.1: Structure of a PBC consisting of two parts: Energy Shaping (ES) and Damping Injection (DI)

6.1.1 Interconnection and damping assignment

IDA-PBC is a specific method combining energy shaping and damping injections described above, and is widely used for the controller synthesis for mechanical systems. The equations from Section 6.1 are adjusted according to the features of mechanical systems to yield a general approach for a class of mechanical systems. The Hamiltonian, or the energy function, in mechanical systems consists, in general, of two parts. The kinetic energy T is given in a quadratic form using the mass matrix $\boldsymbol{M}(\boldsymbol{q})$ and the generalised impulses $\boldsymbol{p}$. The potential energy U usually depends only on the generalised coordinates $\boldsymbol{q}$ resulting in

$$H(\boldsymbol{q},\boldsymbol{p}) = T(\boldsymbol{q},\boldsymbol{p}) + U = \frac{1}{2}\boldsymbol{p}^T \boldsymbol{M}^{-1}(\boldsymbol{q})\,\boldsymbol{p} + U(\boldsymbol{q}). \tag{6.8}$$

It is recommendable to choose the desired Hamiltonian in the same structure by defining a desired mass matrix $\boldsymbol{M}_d(\boldsymbol{q})$ and a desired potential energy $U_d(\boldsymbol{q})$ as design parameter as

$$H_d(\boldsymbol{q},\boldsymbol{p}) = \frac{1}{2}\boldsymbol{p}^T \boldsymbol{M}_d^{-1}(\boldsymbol{q})\,\boldsymbol{p} + U_d(\boldsymbol{q}). \tag{6.9}$$

With that, (6.3) is rewritten as

$$\boldsymbol{M}^{-1}(\boldsymbol{q})\,\boldsymbol{p} = \boldsymbol{J}_1(\boldsymbol{q})\,\boldsymbol{M}_d^{-1}(\boldsymbol{q})\,\boldsymbol{p}, \tag{6.10}$$

which leads to

$$J_1(q) = M^{-1}(q) M_d(q).$$ (6.11)

Due to the symmetry of the mass matrices, one could write

$$-J_1^T(q) = -M_d(q) M^{-1}(q).$$ (6.12)

Substituting (6.11) and (6.12) in (6.4) results in

$$-\frac{\partial^T H(q,p)}{\partial q} + G_u u = -M_d M^{-1} \frac{\partial^T H_d}{\partial q} + J_2 \frac{\partial^T H_d}{\partial p}.$$ (6.13)

For the general case of energy shaping, the control law for a fully actuated system is found to be

$$u_{es} = G_u^{-1}\left(\frac{\partial^T H(q,p)}{\partial q} - M_d M^{-1} \frac{\partial^T H_d}{\partial q} + J_2 \frac{\partial^T H_d}{\partial p}\right).$$ (6.14)

As mentioned before, for an under-actuated system, G_u does not have full rank and therefore, G_u^{-1} does not exist. Using the so-called pseudo-inverse of G

$$G_u^+ = \left(G_u^T G_u\right)^{-1} G_u^T,$$ (6.15)

a control law can be found as

$$u_{es} = G^+\left(\frac{\partial^T H(q,p)}{\partial q} - M_d M^{-1} \frac{\partial^T H_d}{\partial q} + J_2 \frac{\partial^T H_d}{\partial p}\right).$$ (6.16)

However, the resulting input cannot affect all rows of (6.13), and thus, not the entire dynamics of the system, since there exists an annihilator $G_u^{\perp}$ such that

$$G_u^{\perp} G_u = 0.$$ (6.17)

The remaining part of (6.13) results in a set of PDEs which needs to be solved to satisfy (6.13) entirely. The PDE is given by

$$g^{\perp}\left[\frac{\partial^T H(q,p)}{\partial q} - M_d M^{-1} \frac{\partial^T H_d}{\partial q} + J_2 \frac{\partial^T H_d}{\partial p}\right] = 0.$$ (6.18)

Equation (6.18) is called the *restriction equation* and solving it may be complicated for arbitrary nonlinear systems. Solving the restriction equations is therefore often the most challenging step in the synthesis of an IDA-PBC for under-actuated systems. There are already several approaches for specific system classes, for instance for systems with under-actuation of degree 1 such as those from [AOAM05] and [OSGEB02].

6.1.2 Generalised canonical transformation

Generalised Canonical Transformation (GCT) is a particular change of coordinates, used to transform a PH-system into another one, preserving the PH structure [FS01b]. As explained before, GCT is used for designing PBCs since is has deciding advantages. For instance, GCTs can be applied multiple times to a system to achieve different objectives. In fact, every resulting closed loop is a new PH-system to which a new GCT can be applied. This is explained in detail below. Also, using GCT is not constrained to time-invariant systems which makes it proper for tackling trajectory tracking problems.

Assume the time-variant lossless PH-system

$$
\begin{aligned}
\dot{\boldsymbol{x}} &= \boldsymbol{J}(\boldsymbol{x},t)\, \frac{\partial H(\boldsymbol{x},t)}{\partial \boldsymbol{x}}^{T} + \boldsymbol{G}_{\mathrm{pH}}(\boldsymbol{x},t)\, \boldsymbol{u} \\
\boldsymbol{y} &= \boldsymbol{G}_{\mathrm{pH}}^{T}(\boldsymbol{x},t)\, \frac{\partial H(\boldsymbol{x},t)}{\partial \boldsymbol{x}}.
\end{aligned}
\tag{6.19}
$$

A coordinate transformation

$$
\begin{aligned}
\tilde{\boldsymbol{x}} &= \boldsymbol{\Phi}\left(\boldsymbol{x},t\right) \\
\tilde{H} &= H\left(\boldsymbol{x},t\right) + H_{\mathrm{add}}\left(\boldsymbol{x},t\right) \\
\tilde{\boldsymbol{y}} &= \boldsymbol{y} + \boldsymbol{\alpha}\left(\boldsymbol{x},t\right) \\
\tilde{\boldsymbol{u}} &= \boldsymbol{u} + \boldsymbol{\beta}\left(\boldsymbol{x},t\right) \ ,
\end{aligned}
\tag{6.20}
$$

which transforms (6.19) into another PH-system

$$
\begin{aligned}
\dot{\tilde{\boldsymbol{x}}} &= \tilde{\boldsymbol{J}}(\tilde{\boldsymbol{x}},t)\, \frac{\partial \tilde{H}(\tilde{\boldsymbol{x}},t)}{\partial \boldsymbol{x}}^{T} + \tilde{\boldsymbol{G}}_{\mathrm{pH}}(\tilde{\boldsymbol{x}},t)\, \tilde{\boldsymbol{u}} \\
\tilde{\boldsymbol{y}} &= \tilde{\boldsymbol{G}}_{\mathrm{pH}}^{T}(\tilde{\boldsymbol{x}},t)\, \frac{\partial \tilde{H}(\tilde{\boldsymbol{x}},t)}{\partial \boldsymbol{x}} \ ,
\end{aligned}
\tag{6.21}
$$

is called a **generalised canonical transformation**. The Hamiltonian of the transformed system $\tilde{H}$ is changed by adding H_{add}, and $\boldsymbol{\alpha}$ and $\boldsymbol{\beta}$ are the additive output and input transformations, respectively.

For given functions $\boldsymbol{\beta}$ and H_{add} a generalised canonical transformation $\boldsymbol{\Phi}(\boldsymbol{x},t)$ exists, if ([FS01b])

$$
\frac{\partial \boldsymbol{\Phi}}{\partial(\boldsymbol{x},t)} \begin{pmatrix} \boldsymbol{J}\, \dfrac{\partial^{T} H_{\mathrm{add}}}{\partial \boldsymbol{x}} + \boldsymbol{G}_{\mathrm{pH}}\, \boldsymbol{\beta} \\[2mm] -1 \end{pmatrix} = \boldsymbol{0}
\tag{6.22}
$$

Further,

$$
\boldsymbol{\alpha}(\boldsymbol{x},t) = \boldsymbol{G}_{\mathrm{pH}}^{T}\, \frac{\partial^{T} H_{\mathrm{add}}}{\partial \boldsymbol{x}} \ , \quad \tilde{\boldsymbol{J}} = \frac{\partial \boldsymbol{\Phi}}{\partial \boldsymbol{x}}\, (\boldsymbol{J})\, \frac{\partial^{T} \boldsymbol{\Phi}}{\partial \boldsymbol{x}} \quad \text{and} \quad \tilde{\boldsymbol{G}}_{\mathrm{pH}} = \frac{\partial \boldsymbol{\Phi}}{\partial \boldsymbol{x}}\, \boldsymbol{G}_{\mathrm{pH}}.
\tag{6.23}
$$

Figure 6.2 illustrates a GCT graphically.

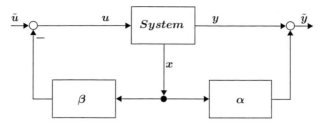

Figure 6.2: Structure of a Generalised Canonical Transformation

Control PH-systems using GCT

Generalised canonical transformations can be used to design a passivity-based controller for PH-system. Assume a general PH-system (2.21) and a generalised canonical transformation (6.20) with $H_{\text{add}} = H_c$ and $\boldsymbol{\beta} = \boldsymbol{\beta}_c$ (Index c for *control*). The new Hamiltonian is

$$\tilde{H}\left(\boldsymbol{q}, \boldsymbol{p}\right) = H(\boldsymbol{q}, \boldsymbol{p}) + H_c. \tag{6.24}$$

The Hamiltonian H_c can be chosen to yield a desired dynamical behaviour of the transformed system, which is the closed loop. Furthermore, if the sufficient and necessary condition

$$\frac{\partial \tilde{H}}{\partial\left(\boldsymbol{x}, t\right)} \begin{pmatrix} \boldsymbol{J}\frac{\partial^T H_c}{\partial \boldsymbol{x}} + \boldsymbol{G}_{\text{pH}}\boldsymbol{\beta}_c \\ -1 \end{pmatrix} \geq 0 \tag{6.25}$$

is satisfied, the resulting system is passive. Inequality (6.25) is also called the *passivity condition* that can be rewritten into

$$\begin{bmatrix} \frac{\partial \tilde{H}(\boldsymbol{q}, \boldsymbol{p})}{\partial \boldsymbol{q}} & \frac{\partial \tilde{H}(\boldsymbol{q}, \boldsymbol{p})}{\partial \boldsymbol{p}} & \frac{\partial \tilde{H}(\boldsymbol{q}, \boldsymbol{p})}{\partial t} \end{bmatrix} \begin{pmatrix} \boldsymbol{J}\frac{\partial^T H_c}{\partial(\boldsymbol{q}, \boldsymbol{p})} + \boldsymbol{G}_{\text{pH}}\boldsymbol{\beta}_c \\ -1 \end{pmatrix} \geq 0. \tag{6.26}$$

The task of stabilising a constant desired equilibrium of a system leads to a time-invariant transformation

$$\frac{\partial \tilde{H}\left(\boldsymbol{q}, \boldsymbol{p}\right)}{\partial t} = 0, \tag{6.27}$$

which simplifies (6.26) to

$$\begin{bmatrix} \frac{\partial \tilde{H}(\boldsymbol{q}, \boldsymbol{p})}{\partial \boldsymbol{q}} & \frac{\partial \tilde{H}(\boldsymbol{q}, \boldsymbol{p})}{\partial \boldsymbol{p}} \end{bmatrix} \begin{pmatrix} \boldsymbol{J}\frac{\partial^T H_c}{\partial(\boldsymbol{q}, \boldsymbol{p})} + \boldsymbol{G}_{\text{pH}}\boldsymbol{\beta}_c \end{pmatrix} \geq 0. \tag{6.28}$$

With a $\boldsymbol{\beta} = \boldsymbol{\beta}_c$ satisfying the condition above, the resulting transformation can be interpreted as energy shaping creating a passive closed loop system with a desired Hamiltonian. Thus, a negative feedback (damping injection) of the output asymptotically stabilises the desired equilibrium

$$\tilde{\boldsymbol{u}} = -\boldsymbol{K}_d\tilde{\boldsymbol{y}} \Rightarrow \boldsymbol{u} + \boldsymbol{\beta}\left(\boldsymbol{x}, t\right) = \boldsymbol{K}_d \cdot \left(\boldsymbol{y} + \boldsymbol{\alpha}\left(\boldsymbol{x}, t\right)\right) \tag{6.29}$$

leading to the overall control law

$$\boldsymbol{u}_{\text{GCT}} = -\boldsymbol{\beta}_c - \boldsymbol{K}_d \cdot \tilde{\boldsymbol{y}} = -\boldsymbol{\beta}_c - \boldsymbol{K}_d \cdot (\boldsymbol{y} + \boldsymbol{\alpha}\,(\boldsymbol{x}, t))\ , \tag{6.30}$$

with $\boldsymbol{\alpha}$ from (6.23).

Integral Extension of GCT

External disturbances or parameter uncertainties in the closed loop may lead to an steady error that is often tackled using an integral action. Within the PBC-framework, this can be achieved by extending the state vector of a PH-system from (2.21) by new states

$$\boldsymbol{x}_I := \int_{t_0}^{t} \boldsymbol{h}^T\,(\boldsymbol{x}, \tau)\,\frac{\partial^T H\,(\boldsymbol{x}, \tau)}{\partial \boldsymbol{x}}\,d\tau\ . \tag{6.31}$$

Note that the matrix $\boldsymbol{h}(\boldsymbol{x}, t)$ is a general parameter that allows for constructing the required integral state $\boldsymbol{x}_I$. For instance, choosing $\boldsymbol{h} = \boldsymbol{I}$ means that $\boldsymbol{x}_I$ is the vector containing the integral of the entire state vector $\boldsymbol{x}$. For a mechanical system with $\boldsymbol{x} = \begin{bmatrix} \boldsymbol{q} & \boldsymbol{p} \end{bmatrix}^T$ however, only the integral of the generalised coordinates $\boldsymbol{q}$ is required that implies choosing $\boldsymbol{h} = \begin{bmatrix} \boldsymbol{I} & \boldsymbol{0} \end{bmatrix}^T$. The resulting overall system is, then, given as

$$\sum\nolimits_{GpHI} : \begin{cases} \dot{\boldsymbol{x}}_e = \left(\begin{bmatrix} \boldsymbol{J} & -\boldsymbol{h} \\ \boldsymbol{h}^T & \boldsymbol{0} \end{bmatrix} - \begin{bmatrix} \boldsymbol{R} & \boldsymbol{0} \\ \boldsymbol{0} & \boldsymbol{0} \end{bmatrix} \right) \frac{\partial^T H}{\partial \boldsymbol{x}_e} + \begin{bmatrix} \boldsymbol{G}_{\text{pH}} \\ \boldsymbol{0} \end{bmatrix} \boldsymbol{u} \\ \boldsymbol{y} = \begin{bmatrix} \boldsymbol{G}_{\text{pH}}^T & \boldsymbol{0} \end{bmatrix} \frac{\partial^T H}{\partial \boldsymbol{x}_e} \end{cases} \tag{6.32}$$

with $\boldsymbol{x}_e = \begin{bmatrix} \boldsymbol{x} & \boldsymbol{x}_I \end{bmatrix}^T$. Note that the structure of (6.32) indicates that the integrator state does not change the original system. Thus, (6.32) represents a passive system in PH-form that is stabilised through the procedure described above. The integral action may also be interpreted as a new GCT applied to the stabilised system by (6.30). For a mechanical system with $\boldsymbol{h} = \begin{bmatrix} \boldsymbol{K}_I & \boldsymbol{0} \end{bmatrix}^T$ for instance, the additional controller input is given by

$$\boldsymbol{u}_I = -\int_{t_0}^{t} \boldsymbol{G}_{\text{pH}} \cdot \boldsymbol{K}_I\,\boldsymbol{q}\,d\tau\ , \tag{6.33}$$

with $\boldsymbol{K}_I = \boldsymbol{K}_I^T > 0$.

6.2 Trajectory tracking control using GCT

The objective of motion control for a nonlinear mechanical system such as a bicycle is to settle an entire desired trajectory and not only a static equilibrium. In general, this requires a non-trivial time-varying controller. Since GCT is a general transformation and not restricted to time-invariant systems, the procedure of controller design can be used for trajectory tracking, too.

The generic approach for trajectory tracking consists of two main steps. First, an error system has to be created with a time varying equilibrium, that is, the deviation from the desired trajectory is zero. In a next step, this equilibrium is stabilised using some controller.

Assume a dynamical system

$$\dot{x} = f(x, u, t) \tag{6.34}$$

and a realisable trajectory $(x^\star, u^\star)$, that is

$$\dot{x}^\star = f(x^\star, u^\star, t) \tag{6.35}$$

holds. A general error system

$$\dot{\bar{x}} = \tilde{f}(\bar{x}, u, t) \quad \text{with} \quad \bar{x}(t) = \Phi(x(t), x^\star(t)) \tag{6.36}$$

is defined to fulfil

$$\bar{x}(t) = 0 \qquad \Leftrightarrow \qquad x(t) = x^\star(t). \tag{6.37}$$

A common approach satisfying (6.37) is

$$\bar{x}(t) = x(t) - x^\star(t). \tag{6.38}$$

Note that (6.36) is, in general, a complex, time-varying system. However, once the transformation is found the problem of trajectory tracking reduces to the stabilisation of the error system.

Transformation into an error system using GCT

Given a PH-system, an error system (6.36) can be created by a generalised canonical transformation. Thereby, the feature of GCT is used which makes sure the transformed system is a passive PH-system.

To achieve the corresponding transformation into the error system (6.22) has to be satisfied. That is a PDE to be solved. There is, however, an approach to find the transformation without solving a PDE which has been introduced for a restricted class of mechanical systems with no gravity, e.g. by [FS02] or [FSS04]. In absence of gravity the potential energy of the system is zero which simplifies the transformation into the error system noticeably. This applies to a range of mechanical systems such as mobile robots or four-wheeled-vehicles on even surfaces. The dynamics of a bicycle, in contrast, strongly depends on the effects of gravity as described in Chapter 4. In this section, an extension of the existing method considering gravity is presented.

Clarification: Please note that in the following the general case of a mechanical PH-system as given by (2.37) is discussed, however with an interconnection matrix in a more general form as

$$J = \begin{bmatrix} 0 & -J_{12}^T \\ J_{12} & J_{22} \end{bmatrix}, \quad J_{22} = J_{22}^T. \tag{6.39}$$

The Hamiltonian is H and the state vector is $x^T = \begin{bmatrix} q & p \end{bmatrix}$. A pseudo-Hamiltonian system, for instance a bicycle that is subject to nonholonomic constraints, can be handled similarly as a special case. In the special case, the system is given by (2.86) with the reduced Hamiltonian $H_\mathcal{M}$ and the state vector $x^T = \begin{bmatrix} q & \rho \end{bmatrix}$. Note also that in this case the mass matrix is replaced by the reduced mass matrix $M_\mathcal{D}$. Furthermore, $J_{12} = S$ and $J_{22} = R$ hold with S and R from (2.86).

Assume the PH-system from (2.37) representing the plant with the Hamiltonian H. Assume also a coordinate transformation

$$\bar{x} = \begin{bmatrix} \bar{q} \\ \bar{p} \end{bmatrix} = \Phi(q, p) = \begin{bmatrix} \Psi \\ \Theta \end{bmatrix}, \tag{6.40}$$

the new Hamiltonian as well as the new input and output

$$\bar{H} = H(q, p) + H_e, \qquad \bar{u} = u + \beta_e, \quad \bar{y} = y + \alpha_e. \tag{6.41}$$

The index $*_e$ stands for the error system. Equation (6.40) is required to

1. build an error system (satisfying (6.37)) and, at the same time,

2. preserve the PH-structure (satisfying (6.22),(6.23) and (6.25)).

Choosing

$$\alpha_e := -\dot{q}^\star, \tag{6.42}$$

with $\dot{q}^\star$ the given desired trajectory for $\dot{q}$, satisfies the first requirement, since $x = x^\star \Rightarrow \bar{y} = 0$ obviously holds. The other way $x = x^\star \Leftarrow \bar{y} = 0$ also holds if the original system is distinguishable, that is for any $t_1 \geq t_0$

$$x(t = 0) = x^\star(t = 0) \text{ and } \bar{y}(t) = 0 \quad \forall t \in [t_0, t_1] \quad \Rightarrow \quad x(t) = x^\star(t) \quad \forall t \in [t_0, t_1]. \tag{6.43}$$

The second requirement indicates that both

$$\frac{\partial^T \Psi(q, t)}{\partial(q, t)} \cdot \begin{bmatrix} J_{12}\alpha \\ -1 \end{bmatrix} = 0 \tag{6.44}$$

and

$$\frac{\partial \bar{H}(q, p)}{\partial(q, p, t)} \left(J \frac{\partial^T H_e(q, p)}{\partial(q, p)} + \begin{bmatrix} 0 \\ G_u \end{bmatrix} \beta_e \right) \geq 0 \tag{6.45}$$

are satisfied. Equation (6.44) is a PDE that depends on the system and has to be solved for every system individually. Note that in case of a system with no nonholonomic constraints $J_{12} = I$ holds that simplifies the solution of (6.44). For those systems, the general solution

$$\Psi(q, t) = \bar{q}(q, q^\star, t) = q - q^\star(t) \tag{6.46}$$

is valid since this implies

$$\left[\begin{array}{cc} \frac{\partial^T \bar{q}}{\partial q} & \frac{\partial^T \bar{q}}{\partial t} \end{array} \right] \cdot \left[\begin{array}{c} J_{12}\alpha_e \\ -1 \end{array} \right] = \left[\begin{array}{cc} I & -\dot{q}^\star \end{array} \right] \cdot \left[\begin{array}{c} -\dot{q}^\star \\ -1 \end{array} \right] = 0. \tag{6.47}$$

For nonholonomic systems such as a bicycle however, this solution is not valid and has to be determined with respect to the corresponding $J_{12} = S$. To make sure that the transformation is a GCT, the additional Hamiltonian H_e has to be chosen such that the first part of (6.23) is satisfied. As an extension to the approach suggested by [FSS04], H_e is chosen to be

$$H_e = \frac{1}{2}\alpha_e^T M(q)\alpha_e + p^T \alpha_e - U. \tag{6.48}$$

This satisfies the output transformation from (6.23) since

$$\frac{\partial H_e}{\partial p} = \alpha_e^T. \tag{6.49}$$

The resulting Hamiltonian is then given as

$$\bar{H} = H(q, p) + H_e = \frac{1}{2}(p + M\alpha_e)^T M^{-1}(p + M\alpha_e). \tag{6.50}$$

The passivity condition (6.45) is considered to obtain the input transformation β_e. Since

$$\frac{\partial H_e}{\partial q} = \frac{1}{2}\alpha_e^T \frac{\partial(M\alpha_e)}{\partial q} - \frac{\partial U}{\partial q}, \tag{6.51}$$

the expression inside the brackets is rewritten into

$$\left[J\frac{\partial H_e}{\partial(q,p)} + \left[\begin{array}{c} 0 \\ G_u \end{array} \right] \beta_e \right] = \left[\begin{array}{c} J_{12}\alpha_e \\ -\frac{1}{2}J_{12}^T\frac{\partial^T(M\alpha_e)}{\partial q}\alpha_e + J_{12}^T\frac{\partial^T U}{\partial q} + J_{22}\alpha_e + G_u\beta_e \\ -1 \end{array} \right]. \tag{6.52}$$

The partial derivatives of $\bar{H}$ with respect to q, p and t are calculated as

$$\frac{\partial \bar{H}}{\partial p} = p^T M^{-1} + \alpha_e^T$$

$$= (p^T + \alpha_e^T M) M^{-1}$$

$$= (p^T + (M\alpha_e)^T) M^{-1}$$

$$= (p + M\alpha_e)^T M^{-1} = \bar{p}^T M^{-1}, \tag{6.53}$$

$$\frac{\partial \bar{H}}{\partial t} = \frac{1}{2} \frac{\partial \left((p + M\alpha_e)^T M^{-1} (p + M\alpha_e) + 2U \right)}{\partial t}$$

$$= \frac{1}{2} \bar{p}^T \frac{\partial \left(M^{-1} \bar{p} \right)}{\partial t} + \left(M^{-1} \bar{p} \right)^T \frac{\partial \bar{p}}{\partial t}$$

$$= \frac{1}{2} \bar{p}^T M^{-1} \frac{\partial \bar{p}}{\partial t} + \bar{p}^T M^{-1} \frac{\partial \bar{p}}{\partial t}$$

$$= \frac{1}{2} \bar{p}^T M^{-1} 2 \left(M \frac{\partial \alpha_e}{\partial t} \right). \tag{6.54}$$

and

$$\frac{\partial \bar{H}}{\partial q} = \frac{1}{2} \frac{\partial \left((p + M\alpha_e)^T M^{-1} (p + M\alpha_e) \right)}{\partial q}$$

$$= \frac{1}{2} \bar{p}^T \frac{\partial \left(M^{-1} \bar{p} \right)}{\partial q} + \left(M^{-1} \bar{p} \right)^T \frac{\partial \bar{p}}{\partial q}$$

$$= \frac{1}{2} \bar{p}^T \frac{\partial \left(M^{-1} \bar{p} \right)}{\partial q} + \bar{p}^T M^{-1} \frac{\partial \bar{p}}{\partial q}$$

$$= \frac{1}{2} \bar{p}^T \left(\frac{\partial \left(M^{-1} p \right)}{\partial q} + M^{-1} \frac{\partial \left(M\alpha_e \right)}{\partial q} \right)$$

$$= \frac{1}{2} \bar{p}^T M^{-1} \left(M \frac{\partial \left(M^{-1} p \right)}{\partial q} + \frac{\partial \left(M\alpha_e \right)}{\partial q} \right). \tag{6.55}$$

Substituting the derivatives above as well as (6.52) into (6.45) and demanding the equation to be valid for all $p(t)$, which removes the term $\bar{p}^T M^{-1}$, the resulting condition is given as

$$\begin{bmatrix} \left(M \frac{\partial \left(M^{-1} p \right)}{\partial q} + \frac{\partial \left(M\alpha_e \right)}{\partial q} \right) \\ I \\ M \frac{\partial \alpha_e}{\partial t} \end{bmatrix}^T \begin{bmatrix} \frac{1}{2} J_{12} \alpha_e \\ -\frac{1}{2} J_{12}^T \frac{\partial^T (M\alpha_e)}{\partial q} \alpha_e + J_{12}^T \frac{\partial^T U}{\partial q} + J_{22} \alpha_e + G_u \beta_e \\ -1 \end{bmatrix} \geq 0. \tag{6.56}$$

Using the equality condition in (6.56), the resulting equation can be solved with respect to β_e delivering a possible input transformation as

$$\beta_e = -\frac{1}{2} G_u^{-1} \left(M \frac{\partial \left(M^{-1} p \right)}{\partial q} + \frac{\partial \left(M\alpha_e \right)}{\partial q} \right) J_{12} \alpha_e + \frac{1}{2} G_u^{-1} J_{12}^T \frac{\partial^T \left(M\alpha_e \right)}{\partial q} \alpha_e \tag{6.57}$$

$$- G_u^{-1} J_{22} \alpha_e - G_u^{-1} J_{12}^T \frac{\partial^T U}{\partial q} + G_u^{-1} M \frac{\partial \alpha_e}{\partial t}.$$

Substituting $\boldsymbol{\beta}_e$ in (6.20), the control input for the transformation into a passive error system with an equilibrium on the desired trajectory is given

$$
\boldsymbol{u}_e = \bar{\boldsymbol{u}} + \frac{1}{2}\boldsymbol{G}_u^{-1}\left(\boldsymbol{M}\frac{\partial\left(\boldsymbol{M}^{-1}\boldsymbol{p}\right)}{\partial\boldsymbol{q}} - \frac{\partial\left(\boldsymbol{M}\boldsymbol{\alpha}_e\right)}{\partial\boldsymbol{q}}\right)\boldsymbol{J}_{12}\boldsymbol{\alpha}_e - \frac{1}{2}\boldsymbol{G}_u^{-1}\boldsymbol{J}_{12}^T\frac{\partial^T\left(\boldsymbol{M}\boldsymbol{\alpha}_e\right)}{\partial\boldsymbol{q}}\boldsymbol{\alpha}_e \quad (6.58)
$$
$$
+ \boldsymbol{G}_u^{-1}\boldsymbol{J}_{22}\boldsymbol{\alpha}_e + \boldsymbol{G}_u^{-1}\boldsymbol{J}_{12}^T\frac{\partial^T U}{\partial\boldsymbol{q}} - \boldsymbol{G}_u^{-1}\boldsymbol{M}\frac{\partial\boldsymbol{\alpha}_e}{\partial t}.
$$

Considering (6.42), (6.58) is rewritten into

$$
\boldsymbol{u}_e = \bar{\boldsymbol{u}} - \frac{1}{2}\boldsymbol{G}_u^{-1}\left(\boldsymbol{M}\frac{\partial\left(\boldsymbol{M}^{-1}\boldsymbol{p}\right)}{\partial\boldsymbol{q}} + \frac{\partial\left(\boldsymbol{M}\dot{\boldsymbol{q}}^\star\right)}{\partial\boldsymbol{q}}\right)\boldsymbol{J}_{12}\dot{\boldsymbol{q}}^\star - \frac{1}{2}\boldsymbol{G}_u^{-1}\boldsymbol{J}_{12}^T\frac{\partial^T\left(\boldsymbol{M}\dot{\boldsymbol{q}}^\star\right)}{\partial\boldsymbol{q}}\dot{\boldsymbol{q}}^\star \quad (6.59)
$$
$$
+ \boldsymbol{G}_u^{-1}\boldsymbol{J}_{22}\dot{\boldsymbol{q}}^\star + \boldsymbol{G}_u^{-1}\boldsymbol{J}_{12}^T\frac{\partial^T U}{\partial\boldsymbol{q}} + \boldsymbol{G}_u^{-1}\boldsymbol{M}\ddot{\boldsymbol{q}}^\star
$$

depending on the time derivative of the desired coordinates $\dot{\boldsymbol{q}}^\star$.

Note that setting the gravity to zero $U = 0$, the general input transformation as from [FSS04] and [TL18a] are recovered, which verifies the generality of the solution. Comparing the result to the so-called method of *virtual potential* from the literature (e.g. [FS02]) for systems without gravity, one may notice the difference to be the term $\boldsymbol{J}_{12}^T\frac{\partial^T U}{\partial\boldsymbol{q}}$. This can be interpreted to be an additional transformation to compensate for the potential energy.

Note also, that the input transformation (6.59) is calculated using the inverse of the input matrix $\boldsymbol{G}_u^{-1}$. As explained before, for an under-actuated system the inverse does not exist and is, therefore, replaced by the pseudo inverse as from (6.15). This requires an additional condition to be satisfied that is explained later.

Using the obtained error system, a controller can be used to stabilize it at the desired trajectory. Here, the stabilisation of the error system is done through a further GCT.

Stabilising the error system

Analogous to Section 6.1.2, the resulting error system is considered a time-varying plant that is stabilised using a GCT with a new Hamiltonian

$$
\tilde{H} = \bar{H} + H_c = H(\boldsymbol{q},\boldsymbol{p}) + H_e + H_c . \quad (6.60)
$$

The corresponding passivity condition (6.25) is given by

$$
\begin{bmatrix} \frac{\partial\tilde{H}(\boldsymbol{q},\boldsymbol{p})}{\partial\boldsymbol{q}} & \frac{\partial\tilde{H}(\boldsymbol{q},\boldsymbol{p})}{\partial\boldsymbol{p}} \end{bmatrix}\left(\boldsymbol{J}\frac{\partial^T H_c}{\partial(\boldsymbol{q},\boldsymbol{p})} + \boldsymbol{G}_u\boldsymbol{\beta}_c\right) \geq 0 , \quad (6.61)
$$

from which the stabilising input $\boldsymbol{\beta}_c$ is calculated as

$$
\boldsymbol{\beta}_c = \boldsymbol{G}_u^{-1}\cdot\boldsymbol{J}_{12}^T\cdot\frac{\partial^T H_c}{\partial\boldsymbol{q}}. \quad (6.62)
$$

Substituting this into (6.20), the control input stabilising the error system is resulted to

$$\bar{u} = \tilde{u} - G_u^{-1} \cdot J_{12}^T \cdot \frac{\partial^T H_c}{\partial q}. \tag{6.63}$$

For asymptotically stabilising the equilibrium of the error system, which is the desired trajectory, a further input as *damping injection* is added leading to the overall control input

$$u = \tilde{u} - \beta_e - \beta_c - K_d \cdot (\dot{r} - \dot{r}_d) \ , \quad K_d = K_d^T > 0 \tag{6.64}$$

where K_d is the damping matrix. Note that $\tilde{u}$ is the input of the resulting transformed system that can be set to zero, a desired input coming out of a trajectory optimisation or be used for a further transformation. Since the equilibrium of the error system is required to be at $\bar{q} = 0$, H_c is chosen a in a well-known quadratic form of

$$H_c = \frac{1}{2}\bar{q}^T K_c \bar{q}. \tag{6.65}$$

6.3 Trajectory tracking control design for two-wheeled vehicles

In this section, the controller explained above is adjusted and applied to two different vehicle models. As explained before, the controller design, especially the solution of the corresponding PDEs, is system specific and has to be derived for every system individually. Therefore, below the controller design is elaborated using a simple system, first, that is the single-track vehicle model derived in Section 4.3. Then, the more complex model of the TWV is tackled.

6.3.1 Trajectory tracking control of a planar single-track vehicle model

In this section, GCT is used to design a trajectory tracking controller for the model of the planar vehicle introduced in Section 4.3. For the derivation of the controller, the *Naive* vehicle described by (4.35) is assumed. However, in the simulations later in this section, the controller is applied to a model containing a positive trail Δ and a non-zero d_s, which act as parameter disturbances. Assuming $d_s = 0$ allows the uncontrolled vehicle to move on a constant circle if started with an initial steering angle while for $d_s \neq 0$ the vehicle tends to move on a straight line. Neglecting the trail Δ changes the steering behaviour of the vehicle, especially under longitudinal acceleration. For instance, accelerating a vehicle with positive trail leads to self-alignment of the steering while an uncontrolled vehicle with $\Delta = 0$ can be accelerated to move faster on a circle. To handle the parameter disturbance, later an adaptive control scheme is introduced.

For the planar vehicle, (6.44) turns into

$$
\begin{bmatrix}
\frac{\partial^T \bar{\xi}}{\partial \xi} & 0 & 0 & 0 & \frac{\partial^T \bar{\xi}}{\partial t} \\
0 & \frac{\partial^T \bar{\sigma}}{\partial \sigma} & 0 & 0 & \frac{\partial^T \bar{\sigma}}{\partial t} \\
0 & 0 & \frac{\partial^T \bar{\eta}}{\partial \eta} & 0 & 0 \\
\frac{\partial^T \bar{\psi}}{\partial \xi} & \frac{\partial^T \bar{\psi}}{\partial \sigma} & 0 & \frac{\partial^T \bar{\psi}}{\partial \psi} & \frac{\partial^T \bar{\psi}}{\partial t}
\end{bmatrix}
\cdot
\begin{bmatrix}
-\dot{\xi}^\star \\
-\dot{\sigma}^\star \\
0 \\
-\sigma \dot{\xi}^\star \\
-1
\end{bmatrix}
= \mathbf{0},
\tag{6.66}
$$

where the first rows are fulfilled choosing

$$
\bar{\xi} = \xi - \xi^\star
$$
$$
\bar{\sigma} = \sigma - \sigma^\star
$$
$$
\bar{\eta} = \eta.
\tag{6.67}
$$

The last row is the PDE

$$
-\frac{\partial^T \bar{\xi}}{\partial \xi} \cdot \dot{\xi}^\star - \frac{\partial^T \bar{\sigma}}{\partial \sigma} \cdot \dot{\sigma}^\star - \frac{\partial^T \bar{\psi}}{\partial \psi} \cdot \sigma \dot{\xi}^\star - \frac{\partial^T \bar{\psi}}{\partial t} = 0
\tag{6.68}
$$

which is satisfied by choosing

$$
\bar{\psi} = 2\left(\psi - \psi^\star\right) - \left(\xi - \xi^\star\right) \cdot \left(\sigma - \sigma^\star\right).
\tag{6.69}
$$

The passivity condition (6.45) is then fulfilled calculating $\boldsymbol{\beta}_e$ as

$$
\boldsymbol{\beta}_e = -\frac{1}{2}\left[\frac{\partial \boldsymbol{M}_D \dot{\boldsymbol{r}}^\star}{\partial \boldsymbol{q}} \boldsymbol{S} - \boldsymbol{M}_D \frac{\partial^T \boldsymbol{M}_D^{-1} \boldsymbol{\rho}}{\partial \boldsymbol{q}} \boldsymbol{S} - \boldsymbol{S}^T \frac{\partial^T \boldsymbol{M}_D \dot{\boldsymbol{r}}^\star}{\partial \boldsymbol{q}} - 2\boldsymbol{R}\right] \dot{\boldsymbol{r}}^\star - \boldsymbol{M}_D \ddot{\boldsymbol{r}}^\star.
\tag{6.70}
$$

The transformation reduces the problem of tracking to a stabilisation problem of the error system, which is approached by a second generalised canonical transformation $\hat{\boldsymbol{x}} = \hat{\boldsymbol{\Phi}}(\bar{\boldsymbol{q}}, \bar{\boldsymbol{p}}, t)$ and the new Hamiltonian $\hat{H} = \bar{H} + \hat{U}$, as explained in Section 6.2. Here, $\hat{U} > 0$ is a virtual potential function having a strict minimum on the trajectory $\bar{\boldsymbol{q}} = \boldsymbol{0}$. Note that the second transformation only adds $\hat{U}$ to the system to stabilise the error system and $\hat{\boldsymbol{x}} = \bar{\boldsymbol{x}}$. Also, $\hat{\boldsymbol{u}} = \bar{\boldsymbol{u}} + \boldsymbol{\beta}_c$ and $\hat{\boldsymbol{y}} = \bar{\boldsymbol{y}}$. Similar to the first transformation, choosing

$$
\boldsymbol{\beta}_c = \boldsymbol{S}^T \frac{\partial \hat{U}}{\partial \bar{\boldsymbol{q}}}, \quad \hat{U} = \frac{1}{2} \bar{\boldsymbol{q}}^T \boldsymbol{K}_c \bar{\boldsymbol{q}}
\tag{6.71}
$$

with $\boldsymbol{K}_c = \boldsymbol{K}_c^T > 0$ renders the closed loop passive with the Hamiltonian $\hat{H}$ which is positive definite with respect to the tracking error. By the negative feedback of the new output to the new input with $\boldsymbol{K}_d > 0$

$$
\hat{\boldsymbol{u}} = -\boldsymbol{K}_d \hat{\boldsymbol{y}} \Rightarrow \boldsymbol{u} = -\boldsymbol{\beta}_e - \boldsymbol{\beta}_c - \boldsymbol{K}_d\left[\boldsymbol{y} - \boldsymbol{y}^\star\right],
\tag{6.72}
$$

the closed loop is asymptotically stable at the desired trajectory. The desired trajectory $\dot{\boldsymbol{r}}^\star$ must be realisable, i.e. it satisfies the dynamics (2.86). Here, a desired path is assumed to be given as functions on the ground plane $x^\star(t)$, $y^\star(t)$. Assuming differentiability, the required states are calculated by

$$\dot{\psi}^\star = \sigma^\star \, \dot{\xi}^\star$$

$$\dot{\xi}^\star = \cos\psi^\star \, \dot{x}^\star + \sin\psi^\star \, \dot{y}^\star$$

$$\dot{\eta}^\star = 0 \;\Rightarrow\; \dot{y}^\star \cos\psi - \dot{x}^\star \sin\psi^\star = 0 \Rightarrow \sigma^\star (\dot{\xi}^\star)^2 = \ddot{y}^\star \cos\psi^\star - \ddot{x}^\star \sin\psi^\star \qquad (6.73)$$

to make the trajectory realisable.

Adaptive extension of the controller

Despite the inherent robustness of the passivity-based controller against parameter uncertainties, large parameter variations can cause undesired behaviour. Here, the controller above is extended by an adaptive scheme to keep the estimation error of some unknown parameter bounded while achieving trajectory tracking. Note that the adaptive extension is a further generalised canonical transformation that preserves the structure of the port-Hamiltonian closed loop.

Assume the unknown vehicle mass m and inertia J with their initial estimation m_0 and J_0 respectively. The reduced mass matrix is written as

$$\begin{aligned}
\boldsymbol{M}_{\mathcal{D}} &= \boldsymbol{M}_{\mathcal{D}0} + \boldsymbol{M}_{\mathcal{D},1}\, z_1 + \boldsymbol{M}_{\mathcal{D},2}\, z_2 \\
&= \begin{bmatrix} J_0\,\sigma^2 + m_0 + m_s & 0 \\ 0 & J_\sigma(\sigma) \end{bmatrix} + \begin{bmatrix} 1 & 0 \\ 0 & 0 \end{bmatrix} z_1 + \begin{bmatrix} \sigma^2 & 0 \\ 0 & 0 \end{bmatrix} z_2
\end{aligned} \qquad (6.74)$$

where z_1 and z_2 are constant and unknown parameter errors with their estimation $\hat{z}_1$ and $\hat{z}_2$ respectively. Note that the estimation errors are defined as $\bar{z}_1 = \hat{z}_1 - z_1$ and $\bar{z}_2 = \hat{z}_2 - z_2$. Rewrite the matrix $\boldsymbol{R}$ as

$$\boldsymbol{R} = \begin{bmatrix} 0 & J\sigma\dot{r}_1 \\ -J\sigma\dot{r}_1 & 0 \end{bmatrix} = \boldsymbol{R}_0 + \boldsymbol{R}_1\, z_1 + \boldsymbol{R}_2\, z_2,$$

$$\boldsymbol{R}_0 = \begin{bmatrix} 0 & J_0\,\sigma\dot{r}_1 \\ -J_0\,\sigma\dot{r}_1 & 0 \end{bmatrix}, \; \boldsymbol{R}_1 = \boldsymbol{0}, \; \boldsymbol{R}_2 = \begin{bmatrix} 0 & \sigma\dot{r}_1 \\ -\sigma\dot{r}_1 & 0 \end{bmatrix} \qquad (6.75)$$

and recall that

$$\boldsymbol{M}_{\mathcal{D}} \frac{\partial (\boldsymbol{M}_{\mathcal{D}}^{-1}\,\boldsymbol{\rho})}{\partial \boldsymbol{q}} = -\frac{\partial (\boldsymbol{M}_{\mathcal{D}}\,\dot{\boldsymbol{r}})}{\partial \boldsymbol{q}}. \qquad (6.76)$$

The control law (6.72) can be written as $\boldsymbol{u} = \boldsymbol{u}_0 + \boldsymbol{\Gamma}\boldsymbol{z}$ with

$$\begin{aligned}
\boldsymbol{u} = &-\boldsymbol{S}^T \boldsymbol{K}_c\,\bar{\boldsymbol{q}} - \boldsymbol{K}_d\,[\dot{\boldsymbol{r}} - \dot{\boldsymbol{r}}^\star] \\
&+ \frac{1}{2}\sum_{l=0}^{2} \hat{z}_l \left[\left(\frac{\partial \boldsymbol{M}_{\mathcal{D},l}\,\dot{\boldsymbol{r}}^\star}{\partial \boldsymbol{q}}\,\boldsymbol{S} + \frac{\partial^T \boldsymbol{M}_{\mathcal{D},l}\,\dot{\boldsymbol{r}}}{\partial \boldsymbol{q}}\,\boldsymbol{S} - \boldsymbol{S}^T \frac{\partial^T \boldsymbol{M}_{\mathcal{D},l}\,\dot{\boldsymbol{r}}^\star}{\partial \boldsymbol{q}} - 2\,\boldsymbol{R}_l \right) \dot{\boldsymbol{r}}^\star + \boldsymbol{M}_{\mathcal{D},l}\,\ddot{\boldsymbol{r}}^\star \right] \qquad (6.77)
\end{aligned}$$

and $z_0 = \hat{z}_0 := 1$. Defining for $w = 1, 2$

$$\boldsymbol{\Gamma}_w := \frac{1}{2} \left(\frac{\partial \boldsymbol{M}_{\mathcal{D},w} \, \dot{\boldsymbol{r}}^\star}{\partial \boldsymbol{q}} \boldsymbol{S} + \frac{\partial^T \boldsymbol{M}_{\mathcal{D},w} \, \dot{\boldsymbol{r}}}{\partial \boldsymbol{q}} \boldsymbol{S} - \boldsymbol{S}^T \frac{\partial^T \boldsymbol{M}_{\mathcal{D},w} \, \dot{\boldsymbol{r}}^\star}{\partial \boldsymbol{q}} - 2 \, \boldsymbol{R}_w \right) \dot{\boldsymbol{r}}^\star + \boldsymbol{M}_{\mathcal{D},w} \, \ddot{\boldsymbol{r}}^\star, \quad (6.78)$$

the adaption law

$$\dot{\hat{z}}_w = -K_{A,w} \boldsymbol{\Gamma}_w^T \dot{\boldsymbol{r}} \tag{6.79}$$

with $K_{A,1}$ and $K_{A,2}$ positive constants results in a new PH system with the extended Hamiltonian

$$H_A = \hat{H} + \frac{1}{2} \bar{\boldsymbol{z}}^T \boldsymbol{K}_A^{-1} \bar{\boldsymbol{z}} \ , \ \boldsymbol{K}_A = \begin{bmatrix} K_{A,1} & 0 \\ 0 & K_{A,2} \end{bmatrix}. \tag{6.80}$$

The resulting closed loop including the parameter estimation is again given in the PH form as

$$\begin{bmatrix} \dot{\hat{q}} \\ \dot{\hat{p}} \\ \dot{\hat{z}} \end{bmatrix} = \begin{bmatrix} \boldsymbol{0} & \hat{\boldsymbol{S}} & \boldsymbol{0} \\ -\hat{\boldsymbol{S}}^T & \hat{\boldsymbol{R}} & \boldsymbol{\Gamma} \, \boldsymbol{K}_A \\ \boldsymbol{0} & -\boldsymbol{K}_A \boldsymbol{\Gamma}^T & \boldsymbol{0} \end{bmatrix} \begin{bmatrix} \frac{\partial^T H_A}{\partial \hat{\boldsymbol{x}}} \\ \frac{\partial^T H_A}{\partial \bar{\boldsymbol{z}}} \end{bmatrix} \tag{6.81}$$

with

$$\begin{bmatrix} \boldsymbol{0} & \hat{\boldsymbol{S}} \\ -\hat{\boldsymbol{S}}^T & \hat{\boldsymbol{R}} \end{bmatrix} = \frac{\partial \boldsymbol{\Phi}}{\partial \boldsymbol{x}} \begin{bmatrix} \boldsymbol{0} & \boldsymbol{S} \\ -\boldsymbol{S}^T & \boldsymbol{R} + \boldsymbol{K}_d \end{bmatrix} \frac{\partial^T \boldsymbol{\Phi}}{\partial \boldsymbol{x}} \bigg|_{\boldsymbol{x} = \boldsymbol{\Phi}^{-1}(\hat{\boldsymbol{x}})} \tag{6.82}$$

that is passive. Thus, it is asymptotically stable at the equilibrium where the tracking error is zero and the estimation error remains unchanged by the negative feedback of the passive output $\hat{\boldsymbol{y}}$ since ([DS12b])

$$\frac{d}{dt} H_A \leq -\hat{\boldsymbol{y}}^T \, \boldsymbol{K}_d \, \hat{\boldsymbol{y}}^T. \tag{6.83}$$

Considering the structure matrix of the closed loop from (6.81), in particular the zero in the (3,3)-component, the estimation error is not fed back as a passive output. Therefore, the estimation error $\bar{\boldsymbol{z}}$ does not converge to zero.

Note that the assumption A.4. from [DS12b] is not fulfilled. However, the assumption A.6. is fulfilled and because of the boundedness of the states in the current case and, thus, the boundedness of $\boldsymbol{\Gamma}_w$, the estimation error becomes constant $\dot{\bar{\boldsymbol{z}}} = \dot{\hat{\boldsymbol{z}}} = \boldsymbol{0}$ and bounded.

Simulation results

To demonstrate the closed loop performance, a horizontal 8 is used as the path given by

$$x^d(t) = 100 \, \cos(\frac{2\,\pi}{100} \, t) \ , \ y^d(t) = 100 \, \sin(\frac{2\,\pi}{50} \, t)$$

Table 6.1: Parameters used for the numerical simulation

Notation	Value	Unit	Short description
m	80	kg	Total mass of the vehicle
m_s	4	kg	Mass of the steering mechanism
J	10	kg m^2	Inertia of the rear frame
J_s	0.1	kg m^2	Inertia of the steering mechanism
l	1.1	m	Wheelbase
l_r	0.4	m	Distance from the rear shaft to the CoM
Δ	5	cm	Trail
d_s	2.5	cm	CoM-shift of the steering mechanism

in combination with (6.73) to generate the desired trajectories. Furthermore, the orientation $\psi(0) = -90°$ is set to make the controlled vehicle have to turn in order to follow the trajectory. For the simulation, the model parameter from the Table 6.1 are used. In the controller (6.77), it is set $\Delta = d_s = 0$, the initial estimation of the vehicle mass is $m_0 = 20kg$ and

$$\mathbf{K}_c = \begin{bmatrix} 15 & 0 & 0 & 0 \\ 0 & 600 & 0 & 150 \\ 0 & 0 & 1 & 0 \\ 0 & 150 & 0 & 20 \end{bmatrix}, \ \mathbf{K}_d = \begin{bmatrix} 60 & 0 \\ 0 & 40 \end{bmatrix}, \ \mathbf{K}_A = \begin{bmatrix} 10 & 0 \\ 0 & 5 \end{bmatrix}. \tag{6.84}$$

The simulation results are illustrated in Figure 6.3, which shows the x-y-plane. The time axis of the state errors $\xi - \xi^\star$ and $\sigma - \sigma^\star$ are shown in Figure 6.4, as well as the estimation for the vehicle mass m. It can be seen, that the state errors approach zero while the parameter estimation becomes constant. Finally to investigate the performance of the controller, a series of 50 simulations (Monte-Carlo simulations) are run with large, randomly varying parameter as given in Table 6.2.

Table 6.2: Parameter variation for investigating the robustness of the closed loop for a single track vehicle model

Parameter	m	l_r	Δ	$\psi(0)$
Nominal value	80	0.4	5	0
Variation range	30 - 200	0.2 - 0.8	0 - 10	-90 - 90
Unit	kg	m	cm	deg

The simulation results are illustrated in Figure 6.5 where the error approaches zero for every variation. In fact, since Δ is chosen to vary between zero and 10 cm, the results indicate that the controller delivers satisfying performance regardless of the unknown positive trail.

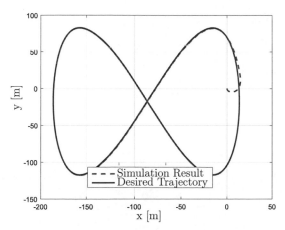

Figure 6.3: Adaptive trajectory tracking control of the planar vehicle: x-y plane

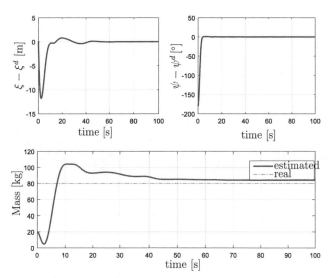

Figure 6.4: Adaptive trajectory tracking control of the planar vehicle: coordinates and the mass estimation

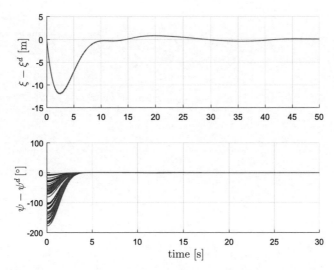

Figure 6.5: Adaptive trajectory tracking control under random variation of parameter and initial conditions

6.3.2 Trajectory tracking control of the *trailed* bicycle model

First, assume a fully actuated model for the bicycle. Basically, the only change is the fact that all three of the generalised forces $\boldsymbol{u}$ are non-zero an can be used in by the controller. This fact is included in the model by setting the input matrix $\boldsymbol{G}_u$ to a unity matrix as

$$
\boldsymbol{G}_{\mathrm{pH}} = \begin{bmatrix} \boldsymbol{0} \\ \boldsymbol{G}_u \end{bmatrix} , \quad \boldsymbol{G}_u = \begin{bmatrix} 1 & 0 & 0 \\ 0 & 1 & 0 \\ 0 & 0 & 1 \end{bmatrix}
\tag{6.85}
$$

The controller design for the more complex bicycle models is based on the same scheme introduced in the Section 6.3.1. The Matrix $\boldsymbol{S}$ for this vehicle model is given as

$$
\boldsymbol{S} = \begin{bmatrix}
1 & 0 & 0 \\
0 & 1 & 0 \\
0 & 0 & 1 \\
\frac{\sin(\delta)}{(L+\Delta)\cos(\delta)-\Delta} & 0 & \frac{\Delta}{(L+\Delta)\cos(\delta)-\Delta} \\
0 & 0 & 0
\end{bmatrix} ,
\tag{6.86}
$$

while the matrix $\boldsymbol{R}$ contains large expressions that are omitted here. All large expressions are included in symbolic and numeric functions from Matlab™ in the digital supplementary to this work.

Similar to the synthesis for the planar vehicle model, condition (6.44) required for construction of the error system turns into

$$
\begin{bmatrix}
\frac{\partial^T \bar{\xi}}{\partial \xi} & 0 & 0 & 0 & 0 & \frac{\partial^T \bar{\xi}}{\partial t} \\
0 & \frac{\partial^T \bar{\varphi}}{\partial \varphi} & 0 & 0 & 0 & \frac{\partial^T \bar{\varphi}}{\partial t} \\
0 & 0 & \frac{\partial^T \bar{\delta}}{\partial \delta} & 0 & 0 & \frac{\partial^T \bar{\delta}}{\partial t} \\
\frac{\partial^T \bar{\psi}}{\partial \xi} & \frac{\partial^T \bar{\psi}}{\partial \varphi} & \frac{\partial^T \bar{\psi}}{\partial \delta} & \frac{\partial^T \bar{\psi}}{\partial \psi} & 0 & \frac{\partial^T \bar{\psi}}{\partial t} \\
0 & 0 & 0 & 0 & \frac{\partial^T \bar{\eta}}{\partial \eta} & 0
\end{bmatrix} \cdot
\begin{bmatrix}
-\dot{\xi}^\star \\
-\dot{\varphi}^\star \\
-\dot{\delta}^\star \\
\frac{-\sin(\delta)\cdot\dot{\xi}^\star - \Delta\cdot\dot{\delta}^\star}{(L+\Delta)\cos(\delta)-\Delta} \\
-1
\end{bmatrix} = \boldsymbol{0}.
\tag{6.87}
$$

Also in this case, the last second row is the challenging PDE to determine $\bar{\psi}$, which is

$$
-\frac{\partial^T \bar{\psi}}{\partial \xi} \cdot \dot{\xi}^\star - \frac{\partial^T \bar{\psi}}{\partial \varphi} \cdot \dot{\varphi}^\star - \frac{\partial^T \bar{\psi}}{\partial \delta} \cdot \dot{\delta}^\star - \frac{\partial^T \bar{\psi}}{\partial \psi} \cdot \left(\frac{-\sin(\delta)\cdot\dot{\xi}^\star - \Delta\cdot\dot{\delta}^\star}{(L+\Delta)\cos(\delta)-\Delta} \right) - \frac{\partial^T \bar{\psi}}{\partial t} - 0.
\tag{6.88}
$$

Two approaches are considered to solve (6.88). The first one is based on the solution introduced in Section 6.3.1 for solving (6.68) that delivers the solution

$$
\bar{\psi} = (\psi - \psi^\star) - \frac{\sin(\delta - \delta^\star)}{(L+\Delta)\cos(\delta - \delta^\star)} \cdot (\xi - \xi^\star)
$$
$$
- \frac{\Delta}{(L+\Delta)\cos(\delta - \delta^\star)} \cdot (\delta - \delta^\star).
\tag{6.89}
$$

The solution above is, however, constrained to the condition

$$\frac{\Delta\dot{\delta}^{\star} + \sin\delta\dot{\xi}^{\star}}{(L + \Delta)\cos(\delta) - \Delta} = 0. \tag{6.90}$$

This constrains the desired trajectories to a circle with $\dot{\delta}^{\star} = 0$ and a real steering angle $\delta = k \cdot \pi$ with $k \in \mathbb{R}$ or a constant longitudinal velocity $\dot{\xi}^{\star} = 0$.

The second approach is, similar to the proposed solution for the planar vehicle (6.69), based on the structure of the Ehresmann connection $\boldsymbol{A}(\boldsymbol{r})$ as

$$\bar{\boldsymbol{s}} = \boldsymbol{s} + \boldsymbol{A}|_{\boldsymbol{r} = \boldsymbol{r}^{\star}}\, \boldsymbol{r}. \tag{6.91}$$

which leads to

$$\bar{\psi} = \frac{\sin\delta^{\star} \cdot \xi + \Delta \cdot \delta}{(L + \Delta)\cos(\delta^{\star}) - \Delta}\,. \tag{6.92}$$

This solution is also constrained to a circular motion, however the condition $\delta = k \cdot \pi$ is no more required which is a relaxation compared to the first approach. This solution was used in the controller synthesis and the simulation below.

Note that the solution satisfying (6.88) only under certain conditions, namely on a circular trajectory, means that the transformation for $\bar{\psi}$ is not valid when the said condition is not fulfilled. In particular, if the vehicle is not on a circular trajectory, $\bar{\psi}$ is not a valid error for the orientation ψ that leads to the controller not maintaining $\psi - \psi^{\star} = 0$. However, since many paths, including a straight line, can be considered as a composition of many circular pieces, the controller is still justified and delivers satisfying closed-loop behaviour. In the simulations below, two sets of trajectories are investigated among which some satisfy the constraining condition and some don't.

Simulation results

For the simulations of the fully actuated bicycle model, different scenarios are considered. It is distinguished between desired trajectories for which (6.92) is a valid solution and those which do not satisfy the constraining condition. For the first case where the desired trajectory is a circular motion, the effect of an input disturbance is investigated. In a second scenario, an arbitrary curved path generated by the vehicle model is considered that violates the constraining condition. Finally in the third scenario, the robustness of the closed loop against parameter variations is investigated where the controller is extended by an integrator according to the section 6.1.2.

Note that all desired trajectories are created using the vehicle model and a simple linear state feedback to guarantee realisability. The state feedback

$$u_{\delta} = -4\,(\varphi - \varphi^{\star}) - 20\,\dot{\varphi} \tag{6.93}$$

where a constant $\varphi^{\star} = 8°$ is used to create circular trajectories and a changing $\varphi^{\star}$ to create the arbitrarily curved path.

Scenario 1: A circular path

In this scenario a circular path is considered with trajectories created using the vehicle model. As external disturbance, impulses on the leaning angle are applied by u_φ which is equivalent to a side-kick to the body of the bicycle. Here, an impulse with amplitude 15 Nm at 50 s and another one with amplitude -15 Nm at 80 s was applied. The matrices $\boldsymbol{K}_c$ and $\boldsymbol{K}_d$ for the control law (6.64) are chosen to be

$$\boldsymbol{K}_c = \begin{bmatrix} 50 & 5 & 0 & 5 & 0 \\ 5 & 200 & 5 & 0 & 0 \\ 0 & 5 & 100 & 0 & 0 \\ 5 & 0 & 0 & 1 & 0 \\ 0 & 0 & 0 & 0 & 1 \end{bmatrix}, \quad \boldsymbol{K}_d = \begin{bmatrix} 50 & 0 & 0 \\ 0 & 70 & 0 \\ 0 & 0 & 150 \end{bmatrix} \tag{6.94}$$

The results are shown in Figure (6.6) where the desired trajectory ($*_{traj}$), the states in absence of external disturbance (no index) and the states subject to external disturbances ($*_{dist}$) are illustrated. The relevant states can be seen which follow the desired trajectory, even despite the external disturbance. Note that the task of the trajectory tracking reduces to maintaining a constant constellation of φ and δ in this case.

The corresponding control inputs are shown in Figure 6.7, where again the inputs in presence and absence of external disturbance are distinguished. It can be seen that the control input has large peaks, especially in u_φ, to compensate for the disturbance, which shall be omitted in a real application.

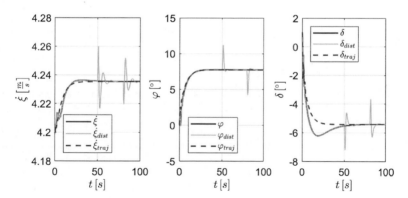

Figure 6.6: Relevant states of a fully actuated bicycle with a circular desired trajectory and subject to side-kick as disturbance

Scenario 2: An arbitrarily curved path

In this scenario, a desired path is investigated which does not satisfy (6.92). Similar to the

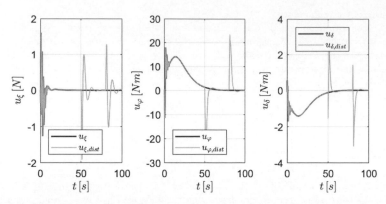

Figure 6.7: Control inputs for a fully actuated bicycle with a circular desired trajectory and subject to side-kick as disturbance

first scenario, the desired trajectory is created using a state feedback and the vehicle model, however with a non-constant $\varphi^\star$ as

$$u_\delta = -\frac{1}{2}\left(\varphi - \varphi^\star(t)\right) - 20\,\dot\varphi\,, \quad \varphi^\star(t) = 8° \cdot \sin\left(\frac{2\pi}{120\,\text{s}} \cdot t\right). \qquad (6.95)$$

For trajectory tracking control, also an integrator was applied as in (6.33). Two simulations were run, one with the integrator involved and one without the integrator setting the matrix $\boldsymbol{K}_I = \boldsymbol{0}$. The controller parameters are chosen to be

$$\boldsymbol{K}_c = \begin{bmatrix} 50 & 0 & 0 & 0 & 0 \\ 0 & 120 & 0 & 0 & 0 \\ 0 & 0 & 120 & 0 & 0 \\ 0 & 0 & 0 & 1 & 0 \\ 0 & 0 & 0 & 0 & 1 \end{bmatrix}, \ \boldsymbol{K}_d = \begin{bmatrix} 300 & 0 & 0 \\ 0 & 250 & 0 \\ 0 & 0 & 200 \end{bmatrix}, \ \boldsymbol{K}_I = \begin{bmatrix} 20 & 0 & 0 \\ 0 & 60 & 0 \\ 0 & 0 & 20 \end{bmatrix}. \qquad (6.96)$$

The simulation results are shown in Figures 6.8 - 6.9. Figure 6.8 illustrates the ground track, however cut in the three time-sections each with a duration of 80 s for better visibility. It can be seen that the desired path is followed with an acceptable accuracy, however not asymptotically due to the violation of the PDE (6.92). Is can also be seen that the integrator reduces the error that is better visible in Figure 6.8, which illustrates the state deviation from the desired trajectory. One may recognise that the resulting deviations are below 10^{-4} m and 10^{-4} °, respectively. Adding the integrator reduces these deviations even below 10^{-4} m and 10^{-4} °. The control inputs are illustrated in Figure 6.10. It is noticeable that adding the integrator not only reduces the state deviation, but also the required input effort.

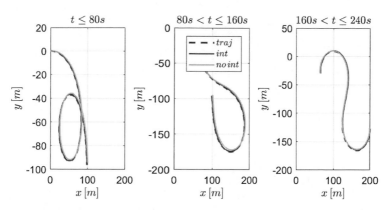

Figure 6.8: Ground track of a fully actuated bicycle for the curved path in xy-coordinates

Scenario 3: Robustness of the closed loop against parameter uncertainties
In the third scenario considered for the fully actuated bicycle model, the robustness of the closed loop against parameter variations is investigated. To this end, a series of Monte-Carlo simulations repeating the first scenario are run, however with randomly varied parameters submitted to the controller. The parameters were randomly chosen within the range given in Table 6.3. Here, $\sim m$ for instance means that the variation range is chosen according to the varied mass m.

Table 6.3: Parameter variation for investigating the robustness of the closed loop for a fully actuated bicycle model

Parameter	m	J	J_x	m_s	J_s	J_{sx}	Δ	h_{CoM}
Nominal value	88	12.4	13	5	0.35	0.42	0.09	0.6
Variation range	50 - 100	$\sim m$	$\sim m$	4-6	$\sim m_s$	$\sim m_s$	0.07 - 0.1	0.5 - 0.7
Unit	kg	kgm^2	kgm^2	kg	kgm^2	kgm^2	m	m

The Monte-Carlos simulations were run both with and without the integrator for comparison. The results for the controller without integrator are illustrated in Figure 6.11. It is visible that the remaining error is small but does not vanish entirely with time. Note that this fact means that the bicycle moves on a different circle as desired, particularly on a circle with a different radius. Simulation results after adding the integrator are shown in Figure 6.12 where, in contrast, the desired angels are tracked asymptotically. The cor-

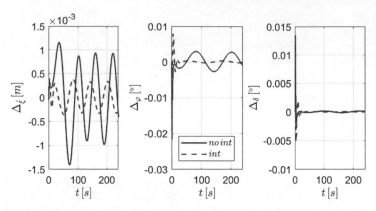

Figure 6.9: State deviations from the trajectory of a fully actuated bicycle for the curved
          path

responding control inputs are shown in figures 6.13 and 6.14 respectively. It can be seen
that the control input does not converge to in case that not integrator is included since the
remaining state errors do not vanish.   In conclusion to scenario 3, circular trajectories
are controlled well using the introduced GCT-based controller. Even for variations in the
system parameters, the error remains bounded. To achieve asymptotic tracking despite
parameter variations, however, an integral action is required which is able to maintain the
goal satisfyingly.

6.3.3 Applying the controller to an under-actuated bicycle model

The controller applied to the bicycle model in Section 6.3.2 delivers evidently satisfying
results. Yet, as mentioned before, it is derived for a fully actuated vehicle model that
assumes a direct torque acting on the leaning angel. Since for a regular TWV this is not
the case, the corresponding vehicle model represents an under-actuated system and, thus,
the bicycle shall be tackled as such. In this section, the introduced controller is adjusted to
handle the under-actuated vehicle model.

For an under-actuated TWV, the input matrix is changed to

$$\boldsymbol{G}_u = \begin{bmatrix} 1 & 0 & 0 \\ 0 & 0 & 0 \\ 0 & 0 & 1 \end{bmatrix}. \tag{6.97}$$

The resulting matrices $\boldsymbol{S}$ and $\boldsymbol{R}$ and, therefore, the rest of the PH system remain unchanged.
There is, however, a severe difference in the control synthesis by GCT due to the passivity

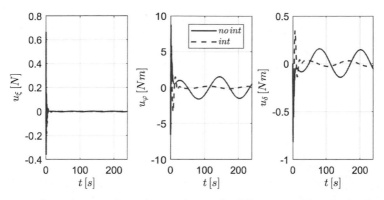

Figure 6.10: Control inputs from the trajectory of a fully actuated bicycle for the curved path

condition (6.25). Similar to IDA for under-actuated systems, for $\boldsymbol{G}_u$ from (6.97) no inverse $\boldsymbol{G}_u^{-1}$ exists. To tackle this issue, a similar approach is used to the one explained in section 6.1.1. Specifically, the inverse matrix in (6.59), (6.63) and (6.33) is replaced by the pseudo-inverse

$$\boldsymbol{G}_u^+ = \left(\boldsymbol{G}_u^T \cdot \boldsymbol{G}_u\right) \cdot \boldsymbol{G}_u^T = \begin{bmatrix} 1 & 0 & 0 \\ 0 & 0 & 0 \\ 0 & 0 & 1 \end{bmatrix}, \tag{6.98}$$

which is equal to the original matrix $\boldsymbol{G}_u^+ = \boldsymbol{G}_u$ in this case. In this particular case and due to the structure of $\boldsymbol{G}_u^+$, this means that the second row of (6.64) corresponding to u_φ is set to zero, or in other words, the second row of the inequality derived from the condition (6.56) is ignored. Since the passivity condition has to be satisfied entirely, one has to make sure additionally that the part of (6.45) that is *not influenced* by the control law, in this case the second row, is fulfilled. Similar to the method explained in Section 6.1.1, this leads to the inequality

$$\boldsymbol{G}_u^\perp \cdot \left[-\frac{1}{2}\left(\boldsymbol{M}_\mathcal{D}\frac{\partial\left(\boldsymbol{M}_\mathcal{D}^{-1}\boldsymbol{\rho}\right)}{\partial\boldsymbol{q}} + \frac{\partial\left(\boldsymbol{M}_\mathcal{D}\dot{\boldsymbol{r}}^\star\right)}{\partial\boldsymbol{q}} \right) \boldsymbol{S}\dot{\boldsymbol{r}}^\star - \frac{1}{2}\boldsymbol{S}^T\frac{\partial^T\left(\boldsymbol{M}_\mathcal{D}\dot{\boldsymbol{r}}^\star\right)}{\partial\boldsymbol{q}}\dot{\boldsymbol{r}}^\star \right.$$
$$\left. -\boldsymbol{R}\dot{\boldsymbol{r}}^\star + \boldsymbol{S}^T\frac{\partial^T U}{\partial\boldsymbol{q}} + \boldsymbol{M}_\mathcal{D}\ddot{\boldsymbol{r}}^\star - \boldsymbol{S}^T\frac{\partial^T H_c}{\partial\boldsymbol{q}} \right] \geq 0, \tag{6.99}$$

where $\boldsymbol{G}_u^\perp$ is an annihilator to $\boldsymbol{G}_u$. Here, $\boldsymbol{G}_u^\perp$ is chosen to be

$$\boldsymbol{G}_u^\perp = \begin{bmatrix} 0 & 1 & 0 \end{bmatrix}. \tag{6.100}$$

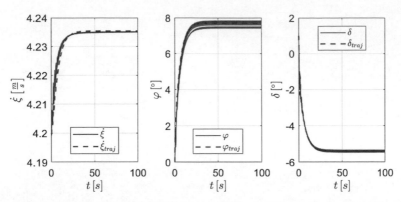

Figure 6.11: Relevant states of a fully actuated bicycle with a circular desired path and subject to parameter variations without integrator

The condition (6.99) is a partial differential inequality for which solutions depend on the system and the desired trajectory. Considering the desired constrained impulse coordinates $\boldsymbol{\rho}^\star(\boldsymbol{r}) := \boldsymbol{M}_\mathcal{D}(\boldsymbol{r})\,\dot{\boldsymbol{r}}^\star$, a rearrangement results in

$$
\boldsymbol{G}_u^\perp \cdot \Bigg[-\frac{1}{2}\left(\boldsymbol{M}_\mathcal{D}\frac{\partial \boldsymbol{r}}{\partial \boldsymbol{q}} + \frac{\partial \boldsymbol{\rho}^\star}{\partial \boldsymbol{q}} \right) \boldsymbol{S}\,\dot{\boldsymbol{r}}^\star - \frac{1}{2}\boldsymbol{S}^T \frac{\partial^T \boldsymbol{\rho}^\star}{\partial \boldsymbol{q}}\,\dot{\boldsymbol{r}}^\star
$$
$$
- \boldsymbol{R}\,\dot{\boldsymbol{r}}^\star + \boldsymbol{S}^T \frac{\partial^T U}{\partial \boldsymbol{q}} + \boldsymbol{M}_\mathcal{D}\,\ddot{\boldsymbol{r}}^\star - \boldsymbol{S}^T \frac{\partial^T H_c}{\partial \boldsymbol{q}} \Bigg] \geq 0\,. \tag{6.101}
$$

Recalling the definition of the matrix $\boldsymbol{S}$ according to (2.83) as $\boldsymbol{S}^T = \begin{bmatrix} \boldsymbol{I} & -\boldsymbol{A}^T \end{bmatrix}$, this equality can be rewritten as

$$
\boldsymbol{G}_u^\perp \cdot \Bigg[-\frac{1}{2}\left(\begin{bmatrix} \boldsymbol{M}_\mathcal{D} & \boldsymbol{0} \end{bmatrix} + \frac{\partial \boldsymbol{\rho}^\star}{\partial \boldsymbol{q}} \right) \begin{bmatrix} \dot{\boldsymbol{r}}^\star \\ -\boldsymbol{A}\,\dot{\boldsymbol{r}}^\star \end{bmatrix} - \frac{1}{2}\begin{bmatrix} \boldsymbol{I} & -\boldsymbol{A}^T \end{bmatrix}\frac{\partial^T \boldsymbol{\rho}^\star}{\partial \boldsymbol{q}}\,\dot{\boldsymbol{r}}^\star
$$
$$
- \boldsymbol{R}\,\dot{\boldsymbol{r}}^\star + \begin{bmatrix} \boldsymbol{I} & -\boldsymbol{A}^T \end{bmatrix}\frac{\partial^T U}{\partial \boldsymbol{q}} + \boldsymbol{M}_\mathcal{D}\,\ddot{\boldsymbol{r}}^\star - \begin{bmatrix} \boldsymbol{I} & -\boldsymbol{A}^T \end{bmatrix}\frac{\partial^T H_c}{\partial \boldsymbol{q}} \Bigg] \geq 0\,. \tag{6.102}
$$

Multiplication of terms yields

$$
\boldsymbol{G}_u^\perp \cdot \Bigg[-\frac{1}{2}\left(\boldsymbol{\rho}^\star + \frac{\partial \boldsymbol{\rho}^\star}{\partial \boldsymbol{r}}\,\dot{\boldsymbol{r}}^\star - \frac{\partial \boldsymbol{\rho}^\star}{\partial \boldsymbol{s}}\boldsymbol{A}\,\dot{\boldsymbol{r}}^\star \right) - \frac{1}{2}\left(\frac{\partial^T \boldsymbol{\rho}^\star}{\partial \boldsymbol{r}}\,\dot{\boldsymbol{r}}^\star - \boldsymbol{A}^T\frac{\partial^T \boldsymbol{\rho}^\star}{\partial \boldsymbol{s}}\,\dot{\boldsymbol{r}}^\star \right)
$$
$$
- \boldsymbol{R}\,\dot{\boldsymbol{r}}^\star + \frac{\partial^T U}{\partial \boldsymbol{r}} - \boldsymbol{A}^T\frac{\partial^T P}{\partial \boldsymbol{s}} + \boldsymbol{M}_\mathcal{D}\,\ddot{\boldsymbol{r}}^\star - \left(\frac{\partial^T H_c}{\partial \boldsymbol{r}} - \boldsymbol{A}^T\frac{\partial^T H_c}{\partial \boldsymbol{s}} \right) \Bigg] \geq 0\,. \tag{6.103}
$$

Since in this particular case

$$
\boldsymbol{G}_u^\perp \cdot \boldsymbol{A}^T = \boldsymbol{0} \tag{6.104}
$$

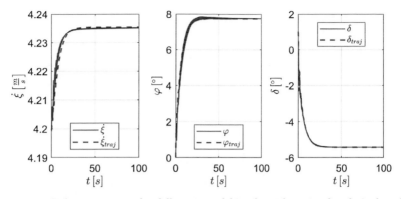

Figure 6.12: Relevant states of a fully actuated bicycle with a circular desired path and subject to parameter variations with integrator

holds and considering (6.100), the inequality reduces to

$$
\begin{aligned}
-\frac{1}{2}\left(\rho_2^\star + 2\frac{\partial^T \rho_2^\star}{\partial r}\,\dot{r}^\star - \frac{\partial^T \rho_2^\star}{\partial s}\begin{bmatrix} A_1^1\,\dot{\xi}^\star + A_3^1\,\dot{\delta}^\star \\ 0 \end{bmatrix}\right) \\
- \boldsymbol{G}_u^\perp\,\boldsymbol{R}\,\dot{r}^\star + \frac{\partial U}{\partial \varphi} + \boldsymbol{G}_u^\perp\,\boldsymbol{M}_\mathcal{D}\,\ddot{r}^\star - \frac{\partial H_c}{\partial \varphi} \geq 0\,,
\end{aligned}
$$
(6.105)

which is rewritten into

$$
\begin{aligned}
-\frac{1}{2}\rho_2^\star + \frac{\partial \rho_2^\star}{\partial \xi}\,\dot{\xi}^\star + \frac{\partial \rho_2^\star}{\partial \varphi}\,\dot{\varphi}^\star + \frac{\partial \rho_2^\star}{\partial \delta}\,\dot{\delta}^\star - \frac{1}{2}\frac{\partial \rho_2^\star}{\partial \psi}\left(A_1^1\,\dot{\xi}^\star + A_3^1\,\dot{\delta}^\star\right) \\
- \boldsymbol{G}_u^\perp\,\boldsymbol{R}\,\dot{r}^\star + \frac{\partial U}{\partial \varphi} + \boldsymbol{G}_u^\perp\,\boldsymbol{M}_\mathcal{D}\,\ddot{r}^\star - \frac{\partial H_c}{\partial \varphi} \geq 0\,.
\end{aligned}
$$
(6.106)

For further simplification, the entries of the following matrices are defined to be

$$
-\boldsymbol{R}\cdot\dot{r}^\star := \boldsymbol{D}^\star = \begin{bmatrix} D_{\dot{\xi}^\star} \\ D_{\dot{\varphi}^\star} \\ D_{\dot{\delta}^\star} \end{bmatrix} \quad \text{and} \quad \boldsymbol{M}_\mathcal{D}\,\ddot{r}^\star := \boldsymbol{F}^\star = \begin{bmatrix} F_{\ddot{\xi}^\star} \\ F_{\ddot{\varphi}^\star} \\ F_{\ddot{\delta}^\star} \end{bmatrix}\,.
$$
(6.107)

With those definitions, the additional condition to make the controller valid results in a compact form as

$$
\begin{aligned}
-\frac{1}{2}\rho_2^\star + \frac{\partial \rho_2^\star}{\partial \xi}\,\dot{\xi}^\star + \frac{\partial \rho_2^\star}{\partial \varphi}\,\dot{\varphi}^\star + \frac{\partial \rho_2^\star}{\partial \delta}\,\dot{\delta}^\star - \frac{1}{2}\frac{\partial \rho_2^\star}{\partial \psi}\left(A_1^1\,\dot{\xi}^\star + A_3^1\,\dot{\delta}^\star\right) \\
+ D_{\dot{\varphi}^\star} + \frac{\partial U}{\partial \varphi} + F_{\ddot{\varphi}^\star} - \frac{\partial H_c}{\partial \varphi} \geq 0.
\end{aligned}
$$
(6.108)

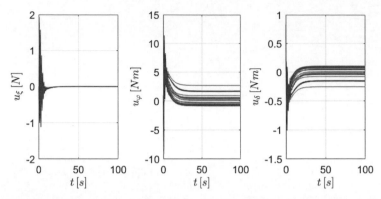

Figure 6.13: Control inputs for a fully actuated bicycle with a circular desired path and subject to parameter variations without integrator

In order to further simplify the condition, note the fact that the partial derivative of the impulse coordinates $\rho^\star$ with respect to the coordinates q depends on the structure of the reduced mass tensor as

$$\frac{\partial \rho_\alpha^\star}{\partial q^i} = \frac{\partial M_{D,\alpha\gamma}(r) \cdot \dot{r}^\gamma}{\partial q^i} = \frac{\partial M_{D,\alpha\gamma}(r)}{\partial q^i} \dot{r}^\gamma. \tag{6.109}$$

Since, however, by definition, the elements of the mass tensor are only functions of the base coordinates r

$$\frac{\partial \rho_\alpha^\star}{\partial s^i} = 0 \ ,$$

holds. This reduces (6.108) to

$$-\frac{1}{2}\rho_2^\star + \frac{\partial \rho_2^\star}{\partial \xi} \dot{\xi}^\star + \frac{\partial \rho_2^\star}{\partial \varphi} \dot{\varphi}^\star + \frac{\partial \rho_2^\star}{\partial \delta} \dot{\delta}^\star + D_{\dot{\varphi}^\star} + \frac{\partial U}{\partial \varphi} + F_{\ddot{\varphi}^\star} - \frac{\partial H_c}{\partial \varphi} \geq 0. \tag{6.110}$$

Furthermore, since the reduced mass tensor in case of the bicycle does not depend on the path coordinate ξ

$$\frac{\partial \rho_\alpha^\star}{\partial \xi} = 0$$

holds reducing the condition further to

$$-\frac{1}{2}\rho_2^\star + \frac{\partial \rho_2^\star}{\partial \varphi} \dot{\varphi}^\star + \frac{\partial \rho_2^\star}{\partial \delta} \dot{\delta}^\star + D_{\dot{\varphi}^\star} + \frac{\partial U}{\partial \varphi} + F_{\ddot{\varphi}^\star} - \frac{\partial H_c}{\partial \varphi} \geq 0. \tag{6.111}$$

A rearrangement yields

$$-\frac{1}{2}\rho_2^\star + \frac{\partial \rho_2^\star}{\partial \varphi} \dot{\varphi}^\star + \frac{\partial \rho_2^\star}{\partial \delta} \dot{\delta}^\star + D_{\dot{\varphi}^\star} + F_{\ddot{\varphi}^\star} + \frac{\partial U}{\partial \varphi} \geq \frac{\partial H_c}{\partial \varphi}. \tag{6.112}$$

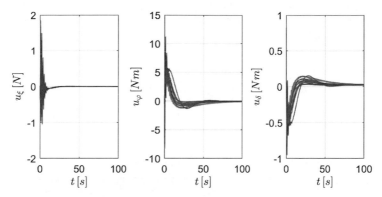

Figure 6.14: Control inputs for a fully actuated bicycle with a circular desired path and subject to parameter variations without integrator

Note that the matrices $\boldsymbol{D}^\star$ and $\boldsymbol{F}^\star$ only depend on the system parameters and the desired trajectory and are not design parameter of the controller. In other words, for given desired trajectories, $\boldsymbol{D}^\star$ and $\boldsymbol{F}^\star$ are assumed to be known matrices. In fact, the only unknown in this inequality is, for given desired trajectories, the partial derivative of the desired Hamiltonian H_c. If for instance, a quadratic function is chosen as suggested in (6.65), the partial derivative is a linear combination of the generalised coordinates as

$$\frac{\partial H_c}{\partial \varphi} = \begin{bmatrix} 0 & 1 & 0 & 0 & 0 \end{bmatrix} \boldsymbol{K}_c \boldsymbol{q}. \tag{6.113}$$

For a diagonal $\boldsymbol{K}_c$, the condition is even further simplified to

$$-\frac{1}{2}\rho_2^\star + \frac{\partial \rho_2^\star}{\partial \varphi}\dot{\varphi}^\star + \frac{\partial \rho_2^\star}{\partial \delta}\dot{\delta}^\star + D_{\dot{\varphi}^\star} + F_{\ddot{\varphi}^\star} + \frac{\partial U}{\partial \varphi} \geq K_{c,22}\,\varphi. \tag{6.114}$$

For an arbitrary function H_c, obviously more complicated expressions are included in the inequality.

Even for a quadratic H_c with a diagonal $\boldsymbol{K}_c$, the inequality (6.114) is still quite complex and finding an explicit solution to it is not trivial. There are however ways to make sure (6.108) is satisfied without finding an explicit solution to the partial differential inequality. Once the desired trajectory is given for instance, one may check in a simulation, if there is a lower bound for the left hand side of (6.114) to choose the value $K_{c,\varphi}$ respectively. In case of the self-stable bicycle, the hypothesis could be issued, that the inequality is satisfied for every, or at least for a certain range of $K_{c,\varphi}$, since (6.114) being satisfied is in some way equivalent to the fact that the bicycle does not fall over while only the steering is being controlled. This is in fact the case as discussed in Chapter 4. In the simulations below, the numeric value of the inequality (6.112) was determined to be positive.

The rest of the control synthesis, for instance the PDE (6.88) is similar to case for the fully actuated vehicle model and remains unchanged.

Simulation results

Simulations for the under-actuated bicycle model were run in the same scheme as for the fully actuated model from Section 6.3.2. Thereby, the same desired trajectories were chosen and the same scenarios were considered.

Scenario 1: A circular trajectory

Similar to the first scenario for the fully actuated vehicle model, two simulations are run, one with no external disturbances, and one with side-kicks on the bicycle frame. The controller is parametrised by the matrices

$$
\boldsymbol{K}_c = \begin{bmatrix} 25 & 10 & 0 & 0 & 0 \\ 10 & 1 & 120 & 0 & 0 \\ 0 & 120 & 60 & 0 & 0 \\ 0 & 0 & 0 & 1 & 0 \\ 0 & 0 & 0 & 0 & 1 \end{bmatrix} , \quad \boldsymbol{K}_d = \begin{bmatrix} 100 & 70 & 0 \\ 70 & 1 & 50 \\ 0 & 50 & 50 \end{bmatrix} . \tag{6.115}
$$

Note that the entry of the matrix $K_{c,23} = K_{c,32}$ is interpreted in the sense that the third state δ is used to affect the third one φ, sine there is no direct actuator acting on the state φ.

The simulation results are shown in Figure 6.15 where the desired leaning angel and the desired circular motion are tracked. The corresponding control inputs are shown in Figure 6.16. Is is noticeable that due to the pseudo-inverse $\boldsymbol{G}_u^+$, the torque on the leaning angle u_φ is eliminated.

Scenario 2: An arbitrarily curved path

Similar to the second scenario for the fully actuated system, in this scenario, an arbitrarily curved path is considered for which the same trajectory is taken as in 6.3.2. Simulations are run once with the controller including the integrator and once without. The controller

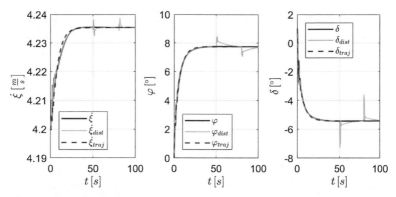

Figure 6.15: Relevant states of an under-actuated bicycle with a circular desired trajectory subject to disturbance

is parametrised using the matrices

$$
\boldsymbol{K}_c = \begin{bmatrix} 10 & 5 & 0 & 0 & 0 \\ 5 & 1 & 120 & 0 & 0 \\ 0 & 120 & 60 & 0 & 0 \\ 0 & 0 & 0 & 1 & 0 \\ 0 & 0 & 0 & 0 & 1 \end{bmatrix} , \boldsymbol{K}_d = \begin{bmatrix} 50 & 50 & 0 \\ 50 & 1 & 50 \\ 0 & 50 & 200 \end{bmatrix} , \boldsymbol{K}_I = \begin{bmatrix} 3 & 20 & 0 \\ 20 & 1 & 20 \\ 0 & 20 & 20 \end{bmatrix} . \quad (6.116)
$$

The results of the ground track are shown in Figure 6.17. Is it visible that the ground path is not tracked asymptotically, however adding the integrator reduces the error noticeably. This can be also observed in Figure 6.18, where the deviation of the relevant states from the desired trajectories are illustrated. The corresponding control inputs are shown in Figure 6.19.

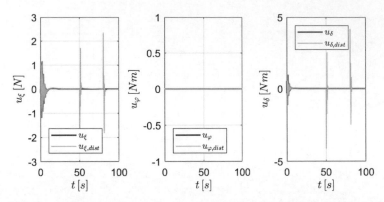

Figure 6.16: Control inputs for an under-actuated bicycle with a circular desired trajectory
subject to disturbance

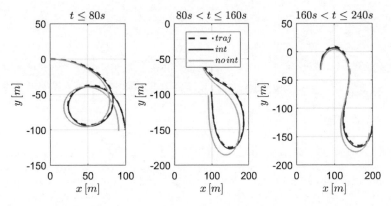

Figure 6.17: Ground track of an under-actuated bicycle for the under-path in xy-coordinates

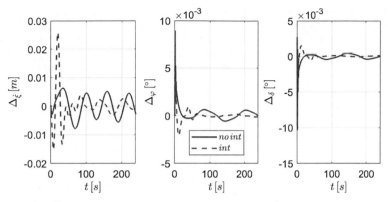

Figure 6.18: State deviations from the trajectory of an under-actuated bicycle for the curved path

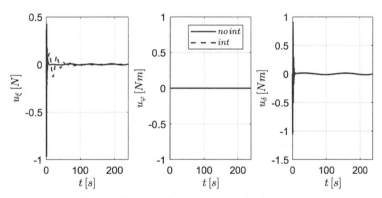

Figure 6.19: Control inputs from the trajectory of an under-actuated bicycle for the curved path

Scenario 3: Robustness of the closed loop against parameter uncertainties

In this scenario, a series of Monte-Carlo simulations is run with randomly varying system parameters to investigate the robustness of the closed loop against parameter uncertainties. In contrast to the case for the fully actuated bicycle model however, here, both the circular path from scenario 1 and the curved path from scenario 2 are investigated. The system parameters are randomly chosen according to the Table 6.3 First, the trajectories corresponding to the circular path are considered, similar to the fully actuated vehicle from 6.3.2. The controller is parametrised as from the first scenario using the same matrices (6.115). Simulation results with a controller without the integral action are illustrated in Figure 6.20, whereas the results from simulations with a controller including integral action are shown in Figure 6.21. It can be observed that the controller with an integrator is able to track the desired trajectory, contrary to he one without integrator. The corresponding control inputs are shown in figures 6.22 and 6.23, respectively.

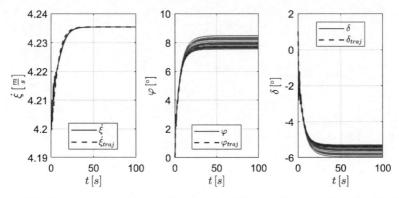

Figure 6.20: Relevant states of an under-actuated bicycle with a circular desired path and subject to parameter variations without integrator

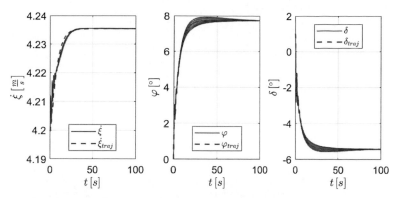

Figure 6.21: Relevant states of an under-actuated bicycle with a circular desired path and subject to parameter variations with integrator

Using the same parameter variation as above, scenario 2 is repeated for the under-actuated vehicle as a series of Monte-Carlo simulations using the same controller parameters from (6.116). Simulation results are illustrated in Figures 6.24 and 6.25 with and without the integrator, respectively. Here again, one can observe that the tracking error on the states is reduced vastly by adding the integrator to the controller. The corresponding control inputs are shown in Figures 6.26 and 6.27, respectively.

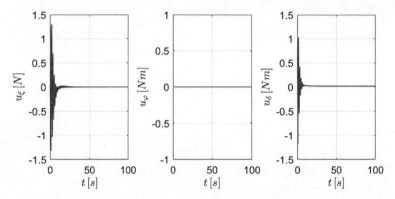

Figure 6.22: Control inputs for an under-actuated bicycle with a circular desired path and subject to parameter variations without integrator

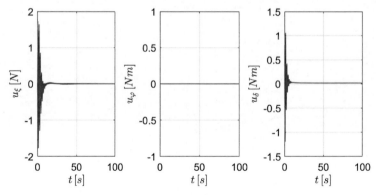

Figure 6.23: Control inputs for an under-actuated bicycle with a circular desired path and subject to parameter variations with integrator

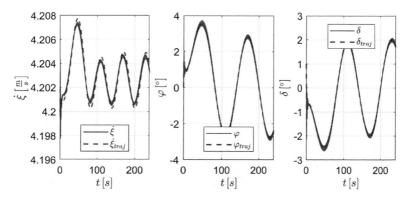

Figure 6.24: Relevant states of an under-actuated bicycle with a curved path and subject to parameter variations without integrator

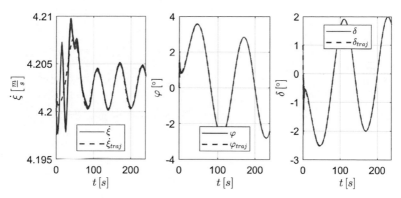

Figure 6.25: Relevant states of an under-actuated bicycle with a curved path and subject to parameter variations with integrator

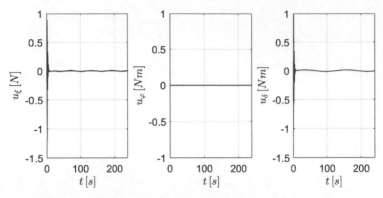

Figure 6.26: Control inputs for an under-actuated bicycle with a curved path and subject to parameter variations without integrator

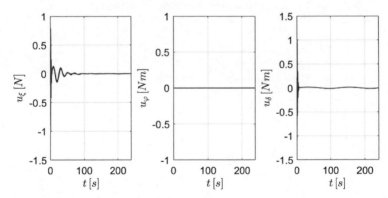

Figure 6.27: Control inputs for an under-actuated bicycle with a curved path and subject to parameter variations with integrator

Comparing the Monte-Carlo simulations for the fully actuated and the under-actuated vehicle model, one may recognize the fact, that the closed loop is more sensitive to parameter variations in the under-actuated case. An integral extension of the controller, however, is able to reduce the tracking error satisfyingly, even for trajectories not fulfilling the PDE (6.88).

In conclusion, the proposed controller is able to track arbitrary realisable trajectories even in presence of parameter uncertainties.

6.3.4 Tracking of the ground path under disturbance

Although the controller from the previous section is very well able to track pre-calculated desired trajectories, tracking a ground path cannot be guaranteed since the error system considers only the generalised coordinates q and not absolute positions of the vehicle on the ground. For instance, the closed loop makes sure that the bicycle moves on a circle with a given radius, however not where the centre of that circle lays in global coordinates. Note that the global coordinates are not part of the considered system, even though they can be calculated during a simulation, e.g. by (4.37). Sticking to the example of a circular path, the initial position of the vehicle decides on the position of the circle in the global coordinates. Furthermore, external disturbances are compensated by the GCT as shown before, yet those generally lead to a shift of the ground path driven by the controlled vehicle.

To deal with this issue and for tracking a-priori determined absolute ground paths, the desired trajectory $q^\star$ has to be *modified*. In other words, another layer has to make sure that the desired coordinates, for instance the desired leaning angle $\varphi^\star$ is changed according to the current state and the desired position on the ground x^p and y^p. Here, this layer is called the *Path Tracking* layer and is implemented as a further controller to create modified desired trajectories. As illustrated in the block diagram from Figure 6.28, it is assumed that the desired path $[x^p, y^p]^T$ and a desired path velocity v^p are given. For the path tracking controller, the idea from [KKMN90] is used that introduces a nonlinear controller for path tracking of nonholonomic robots.

Consider the kinematics of the single-track vehicle

$$\dot{x} = v \cos\psi , \quad \dot{y} = v \sin\psi , \tag{6.117}$$

and define the states x_{kin} and the inputs u_{kin} Assume a given desired path $[x^p, y^p]^T$, a forward velocity v^p and the corresponding values for ψ^p and $\dot{\psi}^p$ creating the vectors

$$x_{kin} := \begin{bmatrix} x \\ y \\ \psi \end{bmatrix} , \quad x_{kin}^p := \begin{bmatrix} x^p \\ y^p \\ \psi^p \end{bmatrix} , \quad u_{kin} := \begin{bmatrix} v \\ \dot{\psi} \end{bmatrix} \tag{6.118}$$

with $v = \dot{\xi}$ the forward velocity. Defining the the *kinematic error system* by the state vector $\tilde{x}_{kin}$ as

$$\tilde{x}_{kin} = \begin{bmatrix} \cos\psi & \sin\psi & 0 \\ -\sin\psi & \cos\psi & 0 \\ 0 & 0 & 1 \end{bmatrix} (x_{kin}^p - x_{kin}) := \begin{bmatrix} \tilde{x} \\ \tilde{y} \\ \tilde{\psi} \end{bmatrix} , \tag{6.119}$$

the control law

$$
\begin{aligned}
v^\star &= v^p \cos\tilde{\psi} + K_x \tilde{x} \\
\dot{\psi}^\star &= \dot{\psi}^p + v^p \left(K_y \tilde{y} + K_\psi \sin\tilde{\psi}\right) \\
K_x,&\ K_y,\ K_\psi > 0
\end{aligned}
\tag{6.120}
$$

asymptotically stabilises the kinematic error system (proof in [KKMN90]). Note that by choosing proper values for K_x, K_y and K_ψ, the controller (6.120) delivers the described modification of the desired trajectory to follow an absolute path.

For using the results by the GCT however, the corresponding leaning angle $\varphi^\star$ and the steering angle $\delta^\star$ need to be determined. Using the steering variable, $\delta^\star$ is determined as

$$
\sigma^\star = \frac{\dot{\psi}^\star}{v^\star}, \quad \delta^\star = \arctan(L\,\sigma^\star)
\tag{6.121}
$$

To obtain $\varphi^\star$, the static balance of the bicycle in a curve is considered. The perpendicular component of the gravity for $\varphi \neq 0$ is compensated by the centripetal acceleration of the CoM in a curve. Mathematically formulated,

$$
g \sin\varphi = -R_c\,\dot{\psi}^2\,\cos\varphi \ ,
\tag{6.122}
$$

where g is the gravity constant and R_c is the radius of the curve. Note that the negative sign is due to the definition of the leaning angle φ and the orientation ψ. Noting that the steering variable is the curvature of the current curve driven by the vehicle $R_c = 1/\sigma$ and considering (6.121), the leaning angle is obtained as

$$
g \sin\varphi = -v\,\dot{\psi}\,\cos\varphi \ .
\tag{6.123}
$$

Finally, (6.121) and (6.123) deliver the required relations

$$
\varphi^\star = -\arctan\left(\frac{v^\star\,\dot{\psi}^\star}{g}\right) \quad \text{and} \quad \delta^\star = \arctan\left(\frac{L\,\dot{\psi}^\star}{v^\star}\right).
\tag{6.124}
$$

In the following simulations, the structure of the control loop is chosen as shown in Figure 6.28. The trajectory tracking controller is chosen to be the GCT using the same parameters as for the curved path from (6.116). Furthermore, the vehicle model is subject to an external disturbance

$$
\boldsymbol{u}_{\text{dist}} =
\begin{bmatrix}
0 \\
u_{\varphi,\text{dist}} \\
0
\end{bmatrix}
$$

with $u_{\varphi,dist}$ simulating a side-kick by a torque impulse every 100 seconds. The side-kick is purposely applied only one-sidedly to challenge the closed loop keeping track of the given path. The path has a closed shape and the desired velocity is chosen to be $v^p = 4.2\,\mathrm{m/s}$. The given path is calculated by choosing a time-varying curvature

$$\sigma^p = \sigma^p_{\max} \sin\left(\frac{2\,\pi}{373.99912}\,t - \frac{\pi}{2}\right) + \frac{1}{500}\,, \quad t \in [0, t_{\mathrm{end}}] \tag{6.125}$$

and integrating the kinematic equations (6.117). The magnitude $\sigma^p_{\max}$ can be used to obtain differently shaped paths. In the simulations, the parameters for the path tracking controller

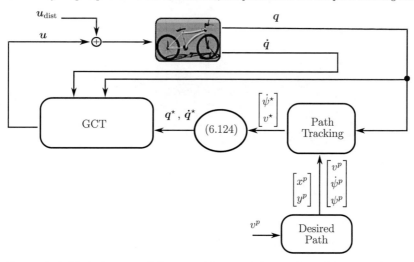

Figure 6.28: Block diagram of the control loop including path tracking controller and GCT

are set to be

$$K_x = 0.01\,, \quad K_y = 0.01\,, \quad K_\psi = 0.1\,. \tag{6.126}$$

For the first simulation, $\sigma^p_{\max} = \frac{1}{500}\,1/\mathrm{m}$ is chosen which leads to an ellipse. The simulation is run with two different settings. First, the path tracking controller is inactive meaning that the desired values for $v^\star$ and $\psi^\star$ are set from the given path equal to v^p and ψ^p, respectively. As illustrated in Figure 6.29, the resulting ground track of the vehicle with no path tracking controller is bent towards the inside due to the one-sided side-kicks although as shown in Figure 6.30 the angles δ and φ converge to the desired values. In contrast, once the path tracking controller is activated, the resulting path follows the desired path. This fact becomes clearer by looking at the states in a zoomed area as illustrated in Figure 6.31. It is visible that the nominal desired angle leaning angle remains constant while the *modified* desired angle obtained by the path tracking controller is changed such that the effect of the external disturbance is compensated for. Further, the simulations are repeated for a more curvy path created by choosing $\sigma^p_{\max} = \frac{1}{250}\,\mathrm{m}$ with the same outcome illustrated in

Figure 6.32. Note that the path tracking controller does not affect the simulations results much when there is no external disturbance involved. This can be observed in the next section, where the trajectory tracking controller is applied for optimal trajectories and no disturbance is applied.

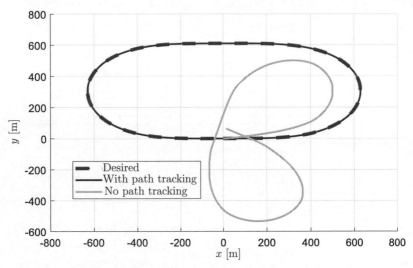

Figure 6.29: Resulting ground track for a controlled under-actuated bicycle with and without additional path tracking controller on an oval desired path

6.3.5 Tracking control for optimal trajectories

As discussed in Section 5.2 and illustrated in Figure 5.2, optimal trajectories obtained by direct collocation consist of grid points (collocation points) and an interpolation between them. For an inherently unstable system such as the TWV model, a good trajectory tracking controller is required to settle the resulting trajectories.

In this final section, the controller from Section 6.3.2 is used to track optimal trajectories created as in Section 5.2. The optimal trajectories are created using the under-actuated *Trailed* model and, thus, the considered scenario is unchanged comparing to that from Section 5.2. The controller from Section 5.2 is applied including the integral action with

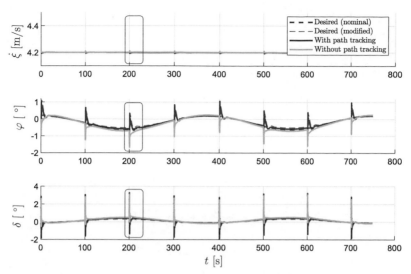

Figure 6.30: Relevant states for a controlled under-actuated bicycle with and without additional path tracking controller on an oval desired path

the following matrices:

$$\boldsymbol{K}_c = \begin{bmatrix} 250 & 150 & 0 & 0 & 0 \\ 150 & 1 & 500 & 0 & 0 \\ 0 & 500 & 10 & 0 & 0 \\ 0 & 0 & 0 & 1 & 0 \\ 0 & 0 & 0 & 0 & 1 \end{bmatrix}, \boldsymbol{K}_d = \begin{bmatrix} 20 & 10 & 0 \\ 10 & 20 & 50 \\ 0 & 50 & 50 \end{bmatrix}, \boldsymbol{K}_I = \begin{bmatrix} 2 & 5 & 0 \\ 5 & 2 & 0 \\ 0 & 0 & 1 \end{bmatrix}. \quad (6.127)$$

Simulation results show that the controller is, as expected, able to track the desired trajectories. One can observe this fact by comparing the simulation results from Figure 6.34 with Figure 5.3, where, most noticeably, the leaning angle φ is tracked very well by the GCT. Furthermore, Figure 6.35 illustrates that the input on the leaning angel u_φ is zero for all t that confirms the tracking for the under-actuated vehicle model. The path on the x-y-plane is shown in Figure 6.33

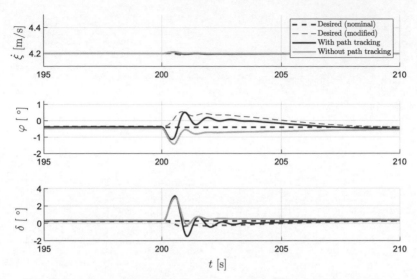

Figure 6.31: Relevant states for a controlled under-actuated bicycle with and without additional path tracking controller on an oval desired path: Zoomed

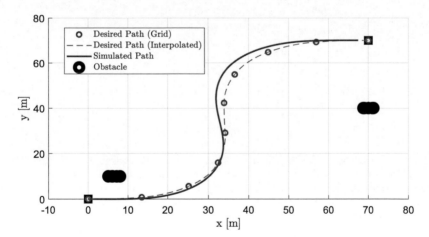

Figure 6.33: Optimal trajectory tracking by GCT on the x-y-plane

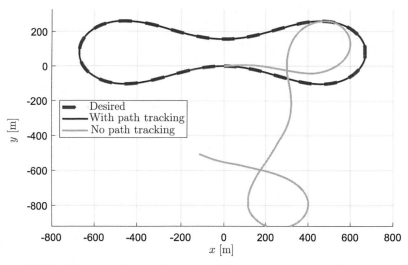

Figure 6.32: Resulting ground track for a controlled under-actuated bicycle with and without additional path tracking controller on an 8-shaped desired path

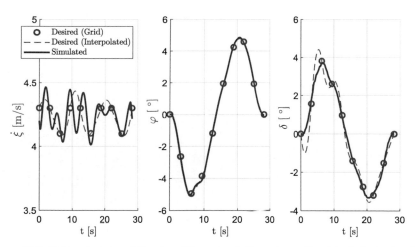

Figure 6.34: Relevant states of the closed loop simulation using optimal trajectories tracked by GCT

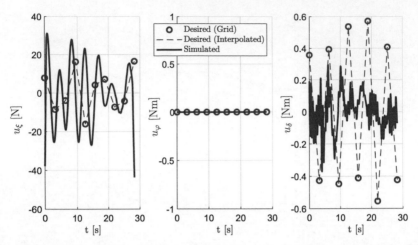

Figure 6.35: Inputs of the closed loop simulation using optimal trajectories tracked by GCT

7 A prototype two-wheeled vehicle for experimental validations

At the Technische Universität Kaiserslautern (TUK), a prototype TWV was developed as an experimental test bench for real-world investigation of the methods involved in autonomous driving of a TWV. The prototype bicycle was entirely designed and developed at the institute for control systems by a team consisting of scientific staff and students with a noticeable support from other units such as technicians and the mechanical workshop. A detailed description of the vehicle, as well as the challenges in the process of the development and operating it, is out of scope of this work. Yet in this chapter, the main components and some functionalities of the prototype bicycle are described briefly.

The bicycle frame is built out of steal profiles which allow easy installation of devices. The main communication bus is Ethernet for Control Automation Technology (EtherCAT). A *Baseline-S* real-time target machine from Speedgoat® was chosen as the main Processing Unit (PU) with *Simulink® Real-Time®* as operating system. In what follows, this machine is loosely referred to as the *Speedgoat*. The *Speedgoat* runs all the motion control algorithms of the bicycle. It is also used as the EtherCAT master for communication with other subsystems slave controllers as well as a data logger via EtherCAT. Also, a state machine is implemented for initialization, safety check, sequential start-up/shutdown of subsystems and management of data from and to the user interface.

To retain the self-stability of the bicycle, a belt system was chosen to connect the steering motor directly (without a gear box) to the steering axis. This helps the steering axis rotate almost freely when the motor is unpowered. Since a high torque is required for this purpose, a Faulhaber® *BLDC 3274G024BP4* is used with nominal values 150 W and a stall torque of 2697 mNm. As driver, Faulhaber® *MC 5010 S ET* Motion Controller is used which has an EtherCAT interface. For the rear wheel propulsion a *BAFANG* FM G31 motor with a rated power of 250 W and maximum torque of 30 Nm was chosen which is commonly used in several E-bikes. The driver was self-developed using the integrated power module *STK984-090A-E*. The control algorithms as well as a special serial communication protocol are hosted on an ATMEGA32.

The perception unit contains different sensors such as Inertial Measurement Unit (IMU) (BNO055) to measure the leaning and orientation angles, a Global Positioning System (GPS) module (NEO 7M-C) and a hall sensor (ENS22F) for accurate measurement of the steering angle as well as its rate. All sensor data is collected on an *Arduino Uno®* and passed to the EtherCAT bus using an *EasyCAT®* shield that provides an EtherCAT slave controller communication to the Arduino via Serial Peripheral Interface (SPI). The User Control Unit (UCU) accepts inputs from the user and sends the gathered input data to the *Speedgaot*. Also, the UCU reads status information from the *Speedgoat* and updates it back to the user. For this purpose, a *Raspberry Pi3®* is used to run a Python script providing a web server on which the user can choose different scenarios and start/stop experiments via WiFi on a front-end device. A small router sets up a local WiFi network. Also, images from the camera installed on the bicycle are streamed to the user vie the same interface.

Up to 10 scenarios are saved as `.dat`-files on *Speedgoat* and can be run once the user chooses one. To run more scenarios, a host computer needs to be connected to *Speedgoat* (Ethernet) to push new files to its memory. A Scenario contains a desired state vector x^*, a desired input vector u^* as well as an *Enable* signal, that is a boolian to mark the beginning and end of the scenario.

Figure 7.1: Picture of the prototype bicycle

Below, in Section 7.1 a short explanation of the Power Distribution and Managemetn Unit (PDMU) is given. Further, the communication system and its various components of it are described. In Section 7.2 the actuation system including the steering actuation and the rear wheel propulsion and the break system are explained. The perception system including the sensors and involved processors is described in Section 7.3. The UCU is described in Section 7.4 and, finally, the software structure of the bicycle control are explained in Section 7.5.

7.1 Overall system architecture

The system architecture needs to facilitate the basic design paradigms so that the system remains simple, flexible and scalable. These three vital design paradigms were considered at the start of the development of the prototype bicycle project and the system architecture was carefully designed to enable future extensions, to reduce conflicts and re-works.

The main components of the prototype vehicle are marked in the picture in Figure 7.2. The mechanical design allows for modular installation of several devices and the steering mechanism is designed with no gears in order to keep the self-stable behaviour of the vehicle as close to reality as possible. The power is supplied by a common e-bike battery. The battery voltage is nominally 36 V which changes from 28 V to 42 V depending on the charge status. However, to adjust the voltage for different electrical components and to avoid failures such as over-voltage, under-voltage, etc. a PDMU is developed that is explained below.

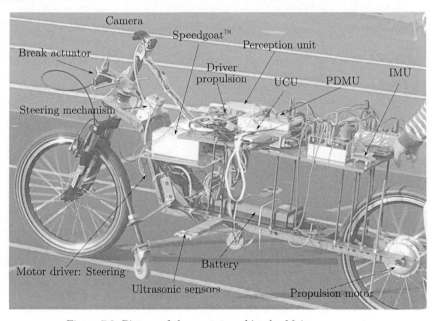

Figure 7.2: Picture of the prototype bicycle: Main components

7.1.1 Power distribution and management

The controllers as well as the motors distributed across the bicycle use different voltage and current levels. Therefore, a main power distribution unit was implemented to provide a separate low level power and high level power supply from the battery. An overview of the power distribution of the system is shown in Figure 7.3. PDMU consists of an Arduino® , a solid-state-relay, measurement circuits and DC/DC converters. The main task of the PDMU is to make sure that the main supply line is cut-off in case a failure occurs. To this end, an signal called *SafeGreen* (5 V) is monitored that is only active once every component sets it actively on high. Among others, the user has to switch this signal manually on. Also, the ultrasonic sensors installed close to the support wheels measure the (adjustable) distance to the ground on every side of the vehicle to set *SafeGreen* to low in case of a fall-over. The low power devices were isolated from the high power devices by using separate DC choppers. A PU typically requires low power levels with voltages ranging from 5V, for instance for the Arduino® and 15 V power supply for the *Speedgoat* and the Faulhaber® driver. On the other hand, the motors require higher power levels. Furthermore, the supply line to the motors are critical and have to be cut off in case of a failure, while the PUs may keep running. For this reason, the power lines are divided into two groups:

- High power lines.
- Low power lines.

Low power lines can be disconnected via manual switches and are directly connected to the battery. The high power lines are connected via the solid-state relay and are only turned on, once the PDMU is in the state *power on*. In case of a failure or by pressing the emergency stop button, the high power lines are disconnected by the relay, yet the low power lines are still active to maintain the functionality of PUs. The bicycle is fitted with a standard battery commonly used in e-bikes. The Lithium Ion battery has the nominal voltage 36 V with a nominal power output of 720 W. The voltage in the fully charged state, however, can rise up to 42 V

An Arduino Uno® is used as PU for the PDMU which measures the voltage of the battery using a measurement circuit and displays it using a 16x2 LCD display. A state machine is implemented consisting of three states that are

- error
- safe mode
- power on

In case of any error or after the initial start-up of the system, the *error* state is active. Once *SafeGreen* is active and after 3 seconds, the state *power on* is active that turns on the solid state relay powering on the high power lines.

7.1.2 Communication of the system

As the main communication bus, EtherCAT was chosen that enables a high data transfer rates up to 200 Megabits per second. In fact, EtherCAT provides many other advantages such as flexibility, possibility to use multiple network topologies and enables Real-Time communication. EtherCAT is an Ethernet based bus system which features standard IEC 61158 real-time network protocol. Thus, the bus system enables soft and hard real-time computing requirements with short cycle time of less than 100 micro seconds, which is essential for the implementation of the system architecture with different controllers distributed across the bicycle. The seamless exchange of data in the bus system lets any PU in the network read and write the data to and from the network bus, thereby eliminating the need for a master device to facilitate internal communication between PUs. The EtherCAT communication requires a master device with at least one standard Network Interface Card (NIC) and the slave device requires an EtherCAT slave controller. All network operation are performed directly by the slave controller, allowing the user to focus only on the controller algorithm.

Since the self-made driver for the rear wheel propulsion motor as well as the UCU are not able to communicate via EtherCAT, those two are connected to the *Speedgoat* via serial communication. An overview of the communication system between the different controller components that are responsible for the individual functions is shown in Figure 7.4.

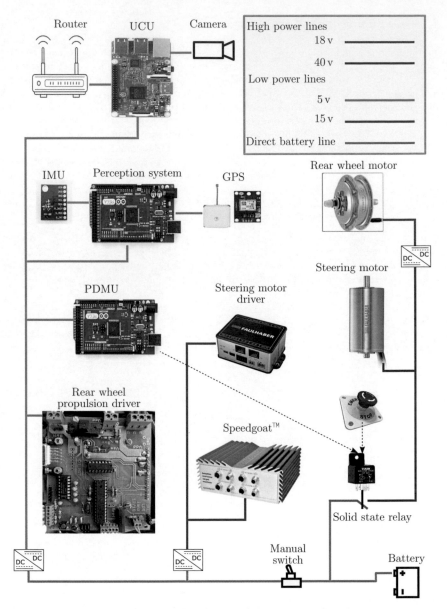

Figure 7.3: Power distribution of the prototype bicycle

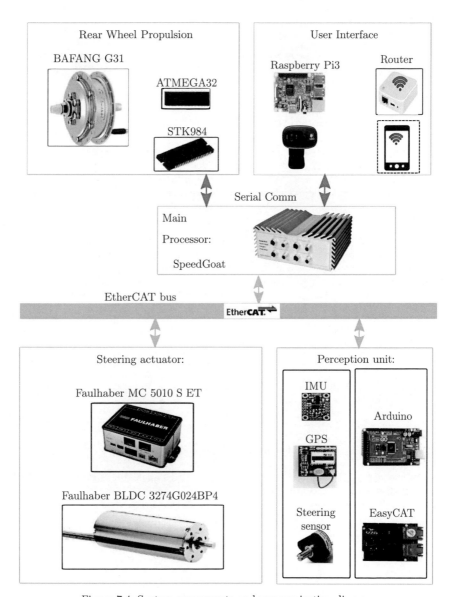

Figure 7.4: System components and communication diagram

7.2 Actuation system

The actuation system consists of two main parts, the propulsion system and the steering actuation. The propulsion system consists of a rear wheel motor and a break system with dedicated actuators for each. In this section, these units are described briefly.

7.2.1 Steering actuation system

The steering actuation system applies required torques on the steering axis of the vehicle. The main components of this system are a Brushless Direct Current (BLDC) motor and the matching Motion Controller (MC), both from Faulhaber® . The Faulhaber® MC is a single axis position controller which controls the speed or position of the supported DC, Brushless and Linear motors. This MC is commonly used for high precision speed control or dynamic position control applications.

The Faulhaber® Motion Controller is available in different series based on the supply voltage and output current requirements as well as in different variants based on the communication interfaces (CANopen, EtherCAT). In accordance to the battery power supply and the chosen communication interface for the system, *MC 5010 S ET* product variant was chosen for the prototype vehicle. The specification of the Drive Electronics *MC 5010 S* is shown in Table 7.1. The BLDC Servomotor *3274G024BP4-3692* (explained in the table 7.2)

Product ID	MC	50	10	S	ET
Info	Motion Controller	Max. supply voltage 50 V	Max. continuous output current 10 A	Housing with screw terminal	EtherCAT interface

Table 7.1: Faulhaber® Motion Contoller Series

from Faulhaber® is a highly efficient, light weight and compact design that delivers high torque. Besides requiring less installation space, the motor is also equipped with an overload protection which makes it suitable for high power applications with dynamic start and stop operation[1]. Some of the important specifications of the motor from the datasheet are provided in the Table 7.3.

---

[1]www.faulhaber.de

Product ID	32	74	G	024	BP4	3692
Info	Diameter 32 mm	Length 74 mm	Series number	Nominal voltage 24 V	4 Pole Variant	Analog Hall sensors

Table 7.2: Faulhaber® Brushless DC-Servomotor Series

Parameters	Values
Power	150 W
Nominal voltage U_N	24 V
Maximum Efficiency η_{max}	89 %
No-load speed n_0	$8700 \, \text{min}^{-1}$
No-load current I_0	0.384 A
Stall torque M_H	2697 mNm
Speed up to n_{max}	$16000 \, \text{min}^{-1}$
Rated torque M_N	162 mNm
Rated current I_N	6.9 A
Rated speed n_N	$8260 \, \text{min}^{-1}$
Hall sensors	Analog

Table 7.3: Faulhaber® Brushless DC-Servomotor Specifications

Motion Controller Operating Modes

The Faulhaber® MC offers several operation modes which are summarised briefly below. The transmission between operation modes as well as activation or deactivation of each mode is done by EtherCAT commanded from the *Speedgoat*.

Homing mode
Before operating the motor in any of the position controller operating modes, a reference run or a reference positioning needs to be performed for the MC. This lets the motor have a reference position based on the application so that the motion controller operates the motor only within the desired range. This reference run or positioning is possible by using the

homing mode of operation. The Homing procedure is generally carried out using a reference switch, limit switch or at current position. The Faulhaber® MC provides several homing modes. One of these modes zeros the position counter at the current position. By this homing method, the straight steering, that is the rear and the front wheel are aligned, is set to the reference position for the motor. This corresponds to a zero steering angle. The homing operation is performed before switching to an operating mode.

Profile position mode

The profile position mode is one of the available modes in the Faulhaber® MC to move the motor to a desired position. Once a target position is communicated to the MC, the profile generator calculates a motion profile for the required position, velocity and acceleration considering maximum values for acceleration, delay and velocity. The target position can either be provided as an absolute position or as a relative position.

Cyclic synchronous torque mode

The cyclic synchronous torque operating mode provides the target torque values for a movement profile in the controller and the drive carries out the torque control operation. This mode was mainly used for the experiments, among others, discussed in Chapter 8.

7.2.2 Rear wheel propulsion and braking system

The rear wheel propulsion and braking system was designed and developed from scratch at the institute for control systems, TUK². This system controls the rear wheel motor speed as well as the front wheel braking actuator. A microcontroller *ATMEGA328* from ATMEL® is used as the PU on which the required c-code is run. The *ATMEGA328* is a 8-bit AVR RISC-based microcontroller which has serial programmable USART, a byte-oriented 2-wire serial interface and an SPI port for external communication. Also, a current sensor *ACS712* is used to measure the input current for the driver.

The system provides a serial RS232 interface to communicate with the *Speedgoat*. The US-ART pins of ATMEGA328 drives the serial communication interface using a *IC MAX232* which adapts the RS-232 signal voltage levels to TTL logic to interface it to a microcontroller.

The desired propulsion or braking actions are achieved by sending data as a custom telegram. Since no standard handshaking signals between the transmitting and receiving sides are used, the telegram is defined with additional bits for proper frame detections.

The microcontroller is interfaced to the rear wheel motor and the stepper motor using a self-made circuit. An intelligent power module *STK984-090A-E* is used as a driver unit

---

²Especially thanks to Mr. Thomas Janz.

for the rear wheel motor and a stepper motor driver *DRV8825* is used for the front brake actuating the stepper motor.

For the rear wheel propulsion a *BAFANG FM G31* motor with a rated power of 250 W and maximum torque of 30 Nm is used. Being small and light weight, it is a highly efficient motor for electric bikes. The 18 V high supplies line powers the rear wheel motor.

A *P542-M481U-G17L82* stepper motor with gear assembly is used to actuate the hydraulic brake of the front wheel motor and a reed switch mounted over the brake is used as position limiter for the stepper motor.

As every other software module running on the *Speedgoat*, the communication with the propulsion driver system is implemented as a Simulink® subsystem. This contains mainly a state machine to determine the health status of the system, a transmission and a receive block. The top layer of this subsystem is shown in Figure 7.5. The wheel rotational

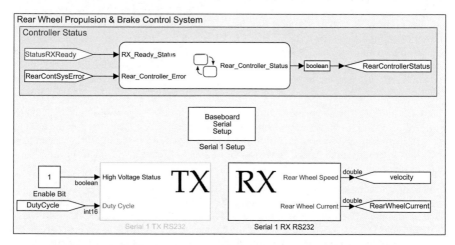

Figure 7.5: Simulink® subsystem for the rear wheel propulsion & brake system

speed and, by the wheel radius, the approximated vehicle velocity is calculated on the microcontroller. This is done using the commutation time that is, then, transmitted to the *Speedgoat*.

7.3 Perception System

The perception system is responsible for collecting the sensor data and sending them to the *Speedgoat*. The sensor data is essential for obtaining the current state vector of the prototype bicycle. To provide real-time characteristics, an Arduino® Mega, a microcontroller board based on the ATmega2560, was chosen. Three Serial ports in the Arduino® Mega make it an ideal option for the perception system to interface different sensors such as IMU and GPS.

The communication between the perception system and the *Speedgoat* is maintained using an EtherCAT slave controller shield EasyCAT® , that is directly connected to the Arduino® Mega. A configuration tool is provided to customise the variables and data types exchanged in EtherCAT and to generate the ESI, header files as well as to flash the EEPROM. Arduino® Mega provides only one SPI to interface which is used for the communication with the EasyCAT® . The angle sensor provides an SPI, too. Thus, the angle sensor is connected to another Arduino® Micro, which receives the data and sends it to the Arduino® Mega over a serial port.

The lean Angle and the lean angular rate are obtained from the Adafruit® *BNO055* absolute orientation sensor. The included Bosch® chip *BNO055* is a System in Package (SIP) which contains a Micro Electro Mechanical System (MEMS), an accelerometer, a magnetometer and a gyroscope along with a ARM® *Cortex-M0* based processor to process the raw data. The calculated quaternions, Euler angles and angular rates are provided over an Inter-Integrated Circuit (I2C) port. The default address for *BNO055* on the I2C bus is 0x28. As two of *BNO055* sensors have been used for better accuracy, the ADR pin of one of the *BNO055* is connected to 3 V to change the I2C address to 0x29.

The Latitude and the Longitudinal position of the bicycle is obtained using a *Waveshare® UART GPS NEO 7M-C*, which is an embedded GPS receiver. It also includes an onboard high-gain active antenna, but for better signal gain an external GPS antenna from Navilock® is used. An Universal Asynchronous Receiver/Transmitter (UART) interface is used for interfacing it to the Arduino® .

By default, the communication protocol of the GPS module is National Marine Electronics Association (NMEA) with a serial baud rate of 9600 bits per second. The module supports seven types of NMEA sentences: $GPGSV, $GPGLL, $GPRMC, $GPVTG, $GPGGA, $GPRMC and $GPGSA. The GPS Parser block provided by Matlab® is used to decode the $GPGGA NMEA data from the GPS Module. Since different types of NMEA sentences are provided by default, the GPS Module is configured to only send the $GPGGA message so that the parser block can decode the position information.

The Angle sensor *ENS22F* is a hall effect rotary encoder from Megatron® . This angle sensor is used to measure the steering angle of the prototype bicycle. The sensor provides an output with a theoretical resolution of 14 bit at an update rate of 200 μs. The effective measured angle of rotation is from 0° to 360° and the mechanical angle of rotation is not limited. This sensor measures the angle using a MLX90316 integrated circuit. The MLX90316 is a Rotary Position Sensor which detects the absolute angular position of a dipole magnet attached to the end of the shaft placed in proximity to the chip. The sensor provides an SPI to read the data from the chip.

The code running on the Arduino® Mega was created by Matlab Coder® which creates C or C++ code from Simulink® models. The corresponding Simulink® model includes a S-function builder for the EtherCAT communication and the blocks to read the sensor data.

The data from the angle sensor is read using the open source libraries - MLX90316 for Serial communication and the Metro library for recurring timed events. The Simulink® model for the Arduino® micro was created to read the angle (in increments) using the SPI interface of the sensor. The function also sends the data through the port *Serial0* of Arduino® Micro to the port *Serial3* of the Arduino® Mega.

The Simulnik subsystem of the perception system on *Speedgoat* maintains the communication with the Arduino® Mega over EtherCAT. Among others, a state machine determines the status of the perception system which depends on the sensors interfaced to it. Based on the values obtained from the sensors, the status of this system is decided to be in failure or OK. This status is then transmitted as a part of the system status to the UCU.

7.4 User Control Unit

The UCU serves as an external interface to the *Speedgoat* in order to provide communication to the prototype vehicle while operating. The UCU accepts inputs from the user and sends the gathered input data to the *Speedgoat*. Also, the control unit reads status information from the system and updates it back to the user.

As only external interface to the vehicle, the UCU has to fulfil the following requirements:

- Communicate to the main controller.

- Read the high power line status input from the PDMU.

- Host a web server for user inputs.

- Set/Reset the status of the *SafeGreen* output.

- Handle a camera stream.

A Raspberry Pi 3® was chosen and used as the PU for the UCU. To meet the requirement to split the critical information such as sensor data from the non-critical logging data over the EtherCAT communication, the existing on-board serial port of the *Speedgoat* was chosen for data exchange with the Raspberry Pi. Therefore, a USB RS232 adaptor was used as the serial communication interface to the UCU.

The *SafeGreen* circuit is used to add a safety signal coming from all sub-systems to control the high power line. All *SafeGreen* circuits are connected in series so that any disruption from any unit disconnects the output from the daisy chain. This circuit has a MEDER® reed relay (SIL05-1A72-71D) and LED for visual indication. Including a *SafeGreen* circuit in the UCU makes sure the user is able to turn off the high power line by a remote command, too.

A Logitech® C270 USB web camera is used to stream the image data to the web interface to serve as a visual reference. Since Flask was used for the hosting a web user interface, the video streaming of the camera to the web user interface was implemented using an *OpenCV* driver with the adapted open source code from the online content.

As a wireless interface for the prototype bicycle, the Raspberry Pi 3® built-in WiFi adapter could be used as a Access Point. But it's evident that the in-built adapter's wireless range is much shorter than an external router such as the TP-Link® 300M Wireless N Nano Router. This router was implemented for a better wireless coverage to access the user web user interface of the autonomous bicycle.

A simple Web User Interface was implemented using Flask to collect the user inputs and to display the camera stream along with the system status. Flask is a microframework for Python based on Jinja 2, which was easy to integrate with the other python functions.

7.5 Software structure

As mentioned before, a Baseline-S Real-time target machine from Speedgoat® - a robust real-time controller with large processing power referred to as the *Speedgoat* - was chosen as the main PU on the prototype bicycle. This rapid prototyping platform is widely used in different real-time application. It also provides different standard input output ports for interfacing sensors and actuators. In the prototype bicycle, it was also used as the EtherCAT master for the communication with the EtherCAT slaves across the vehicle. The Hardware specification of the Baseline-S Real-time target machine is mentioned in Table 7.4.

Configuration	Info
Main Disc	SSD 32GB ExtTemp
CPU	Intel Celeron CPU J1900 @ 1.99GHz
Memory	4GB DDR3 RAM
Installed I/O Modules	No
Baseaddress	0x3F8
Host-Target Port Ethernet Port	1 XCP Slave
EtherCAT Ports	2 XCP Master
Serial Ports	2 x RS232
USB Ports	1 x USB 3.0, 2 x USB 2.0

Table 7.4: *Speedgoat* Hardware Specifications

The main part of the software runs on the *Speedgoat* as a large Simulink® model containing several subsystems. IT is responsible for the coordinating of the perception data, running the control algorithms, transmitting of the control commands to the actuation units as well as maintaining the vehicle safety. Below, some of the important components of the overall software are described briefly.

The Simulink® subsystems included in the main software are shown in Figure 7.6. These subsystems are categorised on a functional and on a system interface level in the target application and are explained in Table 7.5.

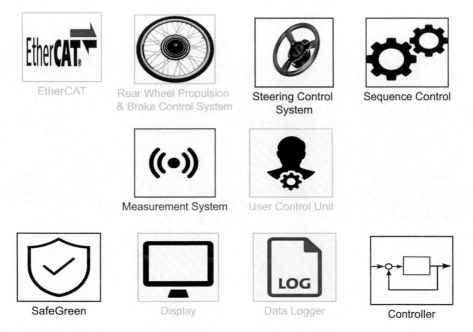

Figure 7.6: Screenshot of the top layer of the Simulink® model run on the *Speedgoat*

Category	Simulink® subsystem	Purpose
Function	EtherCAT	Controls the EtherCAT Master & Slave communication
	Sequence Control	Controls the sequential operation of the bicycle
	Controller	Control algorithms to create actuator commands
	SafeGreen	Safety check of the bicycle
	Display	Target scopes for simulation
	Data Logger	File scopes for data logging
System interface	Rear Wheel Propulsion and Brake Control System	Interface to the driver of the rear wheel and braking
	Steering Control System	Interface to the driver of the steering motor
	Measurement System	Interface to the perception system
	User Control Unit	Interface to communicate the data to the web interface

Table 7.5: Simulink® subsystems - categorised as functional and system interface

7.5.1 EtherCAT communication subsystem

The EtherCAT slaves are controlled based on an EtherCAT state machine. The EtherCAT subsystem includes Simulink® blocks that initialise the EtherCAT Master and controls the state of the EtherCAT network. The subsystem also includes blocks that provide the EtherCAT initialisation error and the EtherCAT network state, which is the EtherCAT network status. This status is further used to activate the *SafeGreen*, and in failure case to deactivate it, which leads to a cut-off of the high power supply lines.

7.5.2 Data logger subsystem

In this subsystem, relevant data is logged and saved during an experiment for post-processing and later investigations. As most of the important data that is supposed to be logged was available only via EtherCAT communication, the data logging functionality was implemented on the *Speedgoat* itself using the Real-Time Scope block from the Simulink® Real-Time library. The data logger is implemented using the file scope and logs the following data which is available on the *Speedgoat*:

- Measured states
- Created inputs
- Desired states
- Desired inputs
- *FromFile* marker

Desired states and desired inputs are created offline within the trajectory planning process. The marker *FromFile* is a bit that is high as long as a desired trajectory is valid. This is used to check the beginning and the end of an experiment.

7.5.3 Display subsystem

The display subsystem includes various scopes which are necessary for monitoring the variables while the application starts executing on the target machine. The Real-Time Scope block is used as a target scope to display the values on a monitor while the vehicle is in the lab and connected to one. Target scopes of numerical type are used for displaying the signal as numerical values, while the graphical rolling is used for displaying the graphical signal.

7.5.4 *SafeGreen* subsystem

The *SafeGreen* Simulink® subsystem controls the *SafeGreen* relay circuit indirectly by communicating the *SafeGreen* status to the UCU. The *SafeGreen* status is either set or reset based on the EtherCAT master status and the steering control system status. This status information is sent to the UCU over serial communication. Based on this status, the *SafeGreen* circuit's relay interfaced to the UCU is turned on or off.

7.5.5 Sequence control subsystem

The sequence control subsystem contains mainly a state machine that determines the current operation mode of the prototype vehicle depending on the internal status of different units as well as the user demand coming through the UCU.

The state machine is provided with inputs from the subsystems *SafeGreen*, UCU and Controller. Based on these inputs, the transitions between the states enables the sequential switch of the operation mode of the vehicle in a desired manner. Apart from these inputs, a software error is also provided to the state machine. A steering angle greater than 61 deg results in software error, which stops the bicycle operation from run to error state. The transition between the states is manually controlled with Sequence Selection input from the user web interface. The sequence control state machine includes the following states in order: Error, Pre-Release, Release, Standby, Run.

Error:
Any undesired condition during the state transitions brings the operation mode to Error state. This state is also used as a reset state to reset the steering driver state machine when switching between different operation modi. The only possible transition from the Error state is the Pre-release state once all the safety conditions are satisfied.

Pre-release:
This additional state in between the Error and the Release provides a waiting period for the *FromFile* marker variable to be zero before starting a new experiment.

Release:
The Release state is mainly used for initialising the vehicle. During Release sequence, the IMU is calibrated and the offsets for the orientation and the lean angle values of the IMU are set. Also the Homing mode of the steering driver is enabled for offsetting the actual position of the steering angle in this sequence. For a free movement of the bicycle the *brake release* command is provided to the Rear Wheel Propulsion and Braking Controller.

Standby:
After the initialisation of the significant parameters, the bicycle can be operated in a standby state, where the steering is held straight and the brake is applied. The steering driver state machine is switched to the Profile Position operating mode to hold the steering straight. A full brake is provided as the duty cycle input to the Rear Wheel Propulsion and Brak

ing Controller. Holding the steering straight with the brakes held is necessary before the feedback controller can take control of the bicycle steering and the rear wheel propulsion.

Run:
The Run state after Standby operation enables the feedback controller and provides the access to control the bicycle steering and the rear wheel propulsion. The feedback controller is only enabled during this sequence.

7.5.6 Measurement subsystem

Here, relevant sensor data is gathered from different communication channels and, if necessary, converted to SI unit values to form the measured state vector as listed in Table 7.6.

Sensor variables	Data	Variable
GPSLatitude	Latitude	x
GPSLongitude	Longitude	y
IMUEulerPitch	Lean angle	φ
SteeringAngle	Steering angle	δ
IMUEulerAzimuth	Orientation	ψ
Speed	Vehicle velocity	v
IMUAngRateX	Lean angle rate	$\dot{\varphi}$
SteeringRate	Steering angle rate	$\dot{\delta}$
IMUAngRateZ	Orientation rate	$\dot{\psi}$

Table 7.6: Measured State vector from sensor data

Position data
 The position of the vehicle on the ground can be obtained in either of the two approaches, which could be chosen based on the user input *EnableGPS* from the web interface of the UCU. In one approach, the relative position is obtained from GPS data relative to a reference position. A Matlab® function calculates the relative position of the bicycle using the latitudinal and longitudinal values from the GPS receiver. In the other approach, the positioning of the bicycle is based on the numeric integration of the kinematics from (4.37).

IMU data
 The Euler angles and their rates obtained by the IMU are given in degree or in degree

per second which are then converted into radians. The IMU Euler azimuth corresponding to the orientation angle ψ is ranged from 0° to 360°. Since the change from 359.99° to 0° in case of a complete turn is undesirable in the orientation data, this angle transformed to range from $-180°$ to 180°. Furthermore, the IMU Euler roll is used to calculate the leaning angle φ and its rate for $\dot{\varphi}$.

Note that the angles ψ and φ do not exactly correspond to the Euler angles coming from the IMU since the chip can hardly be installed perfectly aligned with the bicycle frame. Therefore, after the assembly of the hardware, a calibration delivered a rotation matrix compensating for the installation inaccuracy.

Steering angle and angle rate data
The installed steering sensor delivers the required value for the steering angle. Yet, since the steering angel and its rate are already available in the Faulhaber® driver, these entities are obtained from it over EtherCAT. Furthermore, a manual correction for zero referencing of the angle is no more required in this case since the delivered angle is already offset-free due to the the the homing mode. Since the motor position from the Faulhaber® driver is given in increments, it is converted into angle by a factor of 0.019352333 ($1 = 52 \ increments$). This factor was obtained by comparing the angle obtained from the Angle sensor to the position increments and is used to calculate the measured state.

Speed
The vehicle velocity is approximately calculated by the microcontroller of the rear wheel propulsion unit as mentioned in Section 7.2.2, which was calibrated and verified after the hardware assembly.

7.5.7 Controller subsystem

The controller subsystem shown in Figure 7.7 includes the blocks which are responsible for the implementation of the control algorithms and includes the following subsystems

- **Sensing Unit:** Bundles the measured entities of the measurement subsystem into a measured state vector.

- **Desired State and Input vector:** Reads and unpacks the data for the desired states and inputs coming from the *FromFile* block

- **Feedback Controller:** Here, the main control algorithms are hosted that use the desired state and the measured state to generate actuation commandos

- **Feed-forward Enabler:** This enables the feed-forward path by adding the desired input to the actuation commandos coming from the feedback controller

- **Actuation Unit:** Includes further subsystems that enable or disable actuation input based on the user demand through UCU

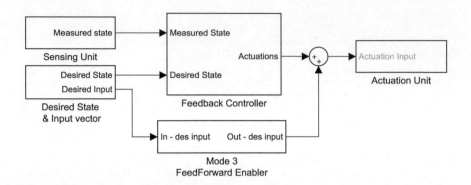

Figure 7.7: Screenshot of the controller subsystem

8 Experimental results

Using the prototype TWV introduced in Chapter 7, some experiments are run to demonstrate the functionality of the prototype vehicle, to investigate the concept of the motion control for TWVs as well as to validate the vehicle model proposed in this work.

There were many challenges that had to be faced even before any experiment can be run. Yet, the primary objective of the prototype development was reached. The main objective of the prototype development and the experiments is to demonstrate a functioning vehicle which is able to ride autonomously while balancing, to exchange data and commands with a remote user as well as to log and save data from experiments. Furthermore, using the prototype vehicle, real-world experiments shall complete the model validation, which was investigated by simulations and by comparison to a benchmark model in Chapter 4.

As explained in Chapter 5, due do the natural instability of a TWV, an investigation of the dynamic behaviour is hardly possible in an open loop. For instance, to examine the validity of the vehicle model by comparing to experimental data, the vehicle has to be controlled to avoid falling over. A controller however, affects the behaviour and the resulting data is falsified. Therefore, an experiment scenario needs to be devised to produce a reasonable comparison between the vehicle model and the experimental results in a closed-loop. The proposed experiments are based on the idea of the simulative investigations in Section 5.2. The setup is similar to the diagram in Figure 5.1, that is, to generate trajectories with different models, and then, to apply them to the most sophisticated model in a closed loop. In this case however, the closed loop contains the actual prototype vehicle instead of a sophisticated model. Furthermore, the trajectories are not created by an optimisation problem, but using a simple state feedback to maintain a desired constant leaning angel, similar to those in Section 6.3.2. This corresponds to a circular ground path. Thereby, the *Trailed* model proposed in Chapter 4 as well as the *Naive* model from the literature are used to create trajectories for riding on a circle with a rather small radius. Similar to the investigations in Section 5.2, the alignment of the reference ground track - namely the circle - with the results of the closed loop is taken as a measure for the quality of the trajectory used in the closed loop. In other words, when the closed loop results in a similar ground track as the simulation for creating the trajectory, that trajectory complies with the vehicle included in the closed loop. Furthermore, the compliance of the created trajectory using a particular vehicle model with the vehicle is a measure for the compliance of that model with the vehicle, and therefore, a measure for the validity of the vehicle model.

This line of thought is used for the experimental model validation in Section 8.2. But first, the functionality of the prototype as well as the motion control for the vehicle using a state feedback is verified experimentally as described in Section 8.1.

The experiments described in this chapter were run on and around the football court of TUK. Some of the experiments were run on the running track that has a flat ground, and some on the grass of the football court. On one hand, the free space on the football court enabled simple run of the experiments. On the other hand, the grass ground caused large disturbances on the measurements of the sensors, which were taken as a challenge. In Figure 8.1 the screenshot of a video is shown which was recorded during the experiments described in Section 8.2 and is available on YouTube® [1].

Figure 8.1: Screenshot of the video illustrating some experiments as well as the location of the experiments on Google® Maps. (GPS: 49.4269556°,7.7529987°)

---

[1]https://youtu.be/d92tF75iQmk

8.1 Experiments for validation of vehicle motion control

To run an experiment, it is worked through the sequence from Section 7.5.5. At the beginning of every experiment, if no error occurred, the release mode is activated. In this mode, the vehicle is moved manually to the position at which the experiment is supposed to start. A file is loaded which contains the desired trajectories for the experiment, by the user selecting a number. Note that every experiment required such a file, even if it only consists of constant or zero values for simple experiments. Furthermore, it is selected whether the feedback controller is applied only, or also the desired inputs are added as a feed-forward. Before running an experiment, it is changed from the release mode to the standby mode. Thereby, the vehicle is held by a user in an upright position and, if required, the sensor data is calibrated. By changing to the run mode, the experiment starts. The experiment stops, when the end of the file containing the desired trajectories is reached. The release mode is automatically activated after the experiment ends and the measurement data is saved on the *Speedgoat*.

As explained in Chapter 7, system states are mainly read directly out of the sensing unit. For the leaning angel φ the pitch-component of Euler-angle provided by the IMU is used. The corresponding rate $\dot{\varphi}$ is directly read from the gyroscope. The steering angel δ and the rate $\dot{\delta}$ is read out of the Faulhaber® motor driver. The forward velocity v is calculated within the rear wheel driver and sent over the serial communication. Note that there are several "jumps" in the measurement of v, especially in the rides on the grass field, as can be seen in Figure 8.2. To determine the states in a better quality, methods of sensor data fusion and filtering are required that is out of the scope of this work. Also the accuracy of the received data from GPS and the corresponding localisation of the vehicle can be increased using correction data and corresponding fusion methods.

To generate the torque command u_δ for the steering motor driver, a linear state feedback is applied as

$$
u_\delta = - \begin{bmatrix} 21.8 & 4.7 & 2.6 & 0.26 \end{bmatrix} \begin{bmatrix} \varphi - \varphi^\star \\ \delta - \delta^\star \\ \dot{\varphi} - \dot{\varphi}^\star \\ \dot{\delta} - \dot{\delta}^\star \end{bmatrix}. \tag{8.1}
$$

These values were chosen heuristically based on trial and error experiments to achieve a good stabilisation of the vehicle at a desired constant leaning angle $\varphi^\star = -8$. For controlling the forward velocity, however, due to inaccuracy in the current measurement, no torque command is generated. Instead, a time-discrete Proportional-Integral (PI)-controller is applied that creates a desired input for the rear wheel driver.

A explained before, desired trajectories are created using the vehicle model in a simulation including a state feedback. For creating the trajectories, the system parameters from the

Table 8.1 are used. Note that these values are partly measured and partly estimated using simplified geometric relations.

Table 8.1: Parameter used for trajectory generation (all SI units)

Parameter	Value	Short description
$\left(J_x, *, J\right)$	$\left(7.16, *, 18.04\right)$	Inertia of rear frame
$\left(J_{sx}, *, J_s\right)$	$\left(0.1071, *, 0.0852\right)$	Inertia of steering
$\left(h, 0, l_r\right)$	$\left(0.45, 0, 0.5\right)$	CoM of rear frame
$\left(h_s, 0, d_s\right)$	$\left(0.6, 0, 0.015\right)$	CoM of steering
m	50	Mass of the rear frame
m_s	9	Mass of the steering
J_w	0.163	Inertia of front wheel
Δ	0.09	Trail
R	0.388	Radius of front wheel
g	9.81	Gravity constant
l	1.25	Wheelbase
ϵ	19°	Head Angle

To prove functionality of the prototype bicycle as well as the ability to ride without falling over, first experiments were run in which the vehicle was supposed to accelerate to a desired constant speed and keep the leaning angle φ as well as the steering angle δ zero. This corresponds to a straight line. Further, trajectories were created for a left-curve and a left-right curve. The left curve corresponds to keeping φ to a constant negative value for a period of time, and the left-right curve to a change of the desired leaning angle from a negative to a positive valued after a while.

The results of some of the experiments are illustrated in Figure 8.3. Note that in Figure 8.4 the south-east corner of the football court was defined as the origin of the world coordinate systems. Red circles mark the starting point of every experiment and green circles mark the end points. Experiment 1 corresponds to a straight line ride. Note that this experiment was repeated often for the start-up of the vehicle, the required modifications and the tuning of the feedback controller (8.1). Experiments 2, 3, 4 and 7 are are left-curves and experiments 5, 6, 8 and 9 left-right curves. The relevant states corresponding to the experiment 6, for instance, are further illustrated in Figure 8.2. Note that on the legends ref denotes the generated desired trajectory by the simulation. `fb` denotes the results by the feedback controller and `ff` the feed-forward input that is zero in this case.

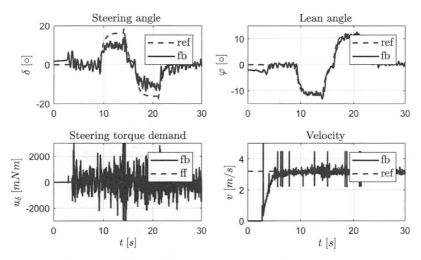

Figure 8.2: Experiment 6: relevant states for a left-right curve

The results demonstrate the functionality of the prototype vehicle as well as the ability of the sate feedback to settle a desired leaning angle. One can see that the leaning angle in Figure 8.2 follows the desired reference curve with a satisfying accuracy. This proves the concept of the motion control for a TWV, since the motion of TWVs is determined basically by the leaning angle, as explained in Section 3.1.

Nevertheless, numerous improvements are considerable. In the experiments in this chapter, a simple linear state feedback is implemented. The results can be improved by the non-linear trajectory tracking controller from Chapter 6. Such a control law, however, requires measured states in a much better quality, especially the rates $\dot{\varphi}$, $\dot{\delta}$ and the speed v. As mentioned before, this requires significantly more development effort that is out of the scope of this work, and, is matter of the future work. Furthermore, improving the quality of the vehicle localisation, the controller architecture from Section 6.3.4 can be used to complete the task of autonomous driving by adding automated navigation.

8.2 Experiments for validation of the vehicle model

As explained above, for the experimental validation of the proposed model, a scenario is considered, in which trajectories corresponding to a circular ground track are compared.

For creating the trajectories, simulations are run using the *Trailed* model as well as the *Naive* model with a state feedback to settle a constant desired leaning angle of $-8°$, such that the resulting reference path circle with a radius of roughly 7.5 m.

Four different variations of the same experiment were run distinguished by

1. whether the desired inputs $u_\delta^\star$, $u_\xi^\star$ and the desired states $\boldsymbol{x}^\star$ are generated by the *Naive* or the *Trailed* model, and

2. whether the system inputs are calculated only by a feed-back of the desired states, or the desired inputs are added as a feed-forward.

Every experiment was repeated at least three times to make sure that the recorded data and the resulting conclusions are consistent. As shown in Figure 8.1, for consistency, every experiment started from the same point with the bicycle - as far as possible - in the same initial orientation. Figure 8.1 also includes 4 shots of the various videos[2] which were recorded during the experiments.

The resulting ground tracks are shown in Figure 8.4. The leaning angle φ, steering angle δ, forward velocity v as well as the demanded input torque on the steering angle u_δ are shown in Figure 8.5 for the trajectories created by the *Naive* model, and in Figure 8.6 for the trajectories created by the *Trailed* model. The starting point is marked as a red point. The radius of the desired path is chosen intentionally small to emphasise the distinction between the two models. In this way, a smaller circle as a resulting ground track of the experiment indicates a better trajectory in the sense explained above. One can see that both ground tracks with and without feed-forward corresponding to the *Naive* model, `Naive:fb+ff` and `Naive:fb` respectively, result in larger circles comparing to those corresponding to the *Trailed* model , `Trailed:fb+ff` and `Trailed:fb` respectively. Adding feed-forward improves the tracking of the leaning angle in each case, and leads therefore to the a smaller circle. The relevant states δ, φ, v and the input u_δ are illustrated in Figures 8.6 and 8.5. One may notice that the change of the leaning angel as well as its tracking is much more natural in case of the trajectories by the *Trailed* model. This leads also to a higher torque command from the feedback. This fact becomes clearer looking at the area around 10 s, where the leaning angle in Figure 8.6 changes much smoother to the desired value than in Figure 8.5. The same effect can be observed for the steering angle which has a smoother transition in Figure 8.6. One explanation can be that the demanded steering angle in the trajectory (`ref`) having a large peak for the counter-steering of almost $-12°$ in the trajectory by the *Naive* model. This leads to the circular ride starting later and, therefore, having their centre further in the south compared to the case with the trajectories by the *Trailed* model as shown in Figure 8.4.

---

[2]Video available at `https://youtu.be/d92tF75iQmk`. From the minute 2:13 on the experiments illustrated in Fig. 8.4 are played.

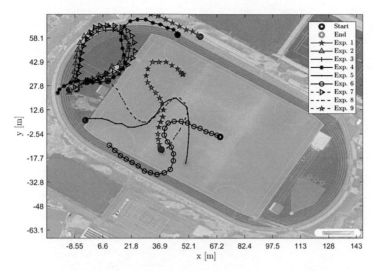

Figure 8.3: Different experiments: Ground Tracks (Google® Maps)

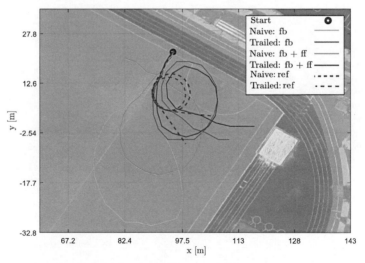

Figure 8.4: Circle rides at the corresponding location on Google®

It is obvious that in none of the experiments the small reference circle of the ground track is followed by the vehicle, which was not the intension of the experiments. The experiments mainly indicate that the *Trailed* model represents the behaviour of a TWV better than the *Naive* model, that is the conclusion of this section.

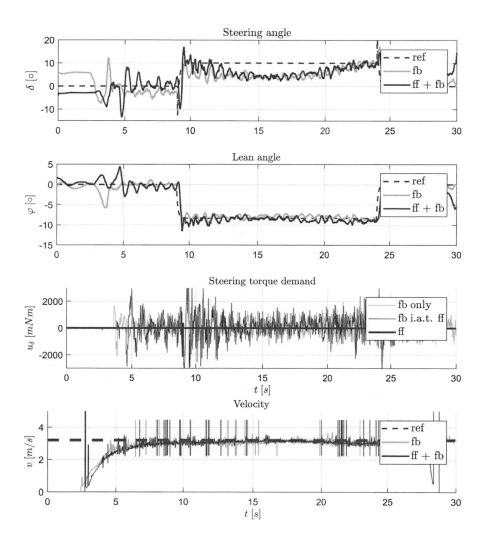

Figure 8.5: Results for the trajectories created by the **Naive** model

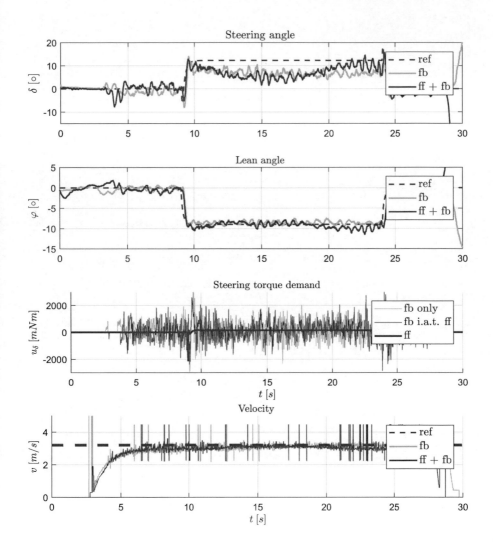

Figure 8.6: Results for the trajectories created by the *Trailed* model

9 Conclusion and future work

Conclusion

With respect to the future urban mobility, modern electrical bicycles, advanced motorcycles and innovative two-wheeled vehicles are inevitable participants of the highly automated and networked traffic system. The vision to be picked up by an autonomous two-wheeled vehicle at a bus station for the last mile to the office is as realistic as an autonomous two-wheeled vehicle responsible for parcel delivery. Dynamical behaviour of two-wheeled vehicles, especially the so-called self-stability, has been studied since decades. There exist a variety of scientific literature with theoretical and experimental investigations of the effect of different parameters and their changes on the self-stability of two-wheeled vehicles. Yet, even today there is no confirmed answer to the question: why exactly are moving bicycles self-stable in a certain velocity range?

Many existing models for bicycles are based on extensive investigations of the physics which are, however, not usable for model-based controller design. On the other hand, there exist models in a proper structure for controller design using well-known synthesis methods, which are often too simplified to cover the interesting physical behaviour such as the self-stability. In this thesis, a systematic method is introduced to unify the approaches to modelling and model-based controller design for autonomous two-wheeled vehicles.

Using the Lagrangian and Hamiltonian frameworks, a sophisticated vehicle model is developed representing important physical phenomenon involved in the behaviour of two-wheeled vehicles. The model is verified by a series of simulation scenarios, as well as by a comparison to available bicycle models from the literature. The systematic extendibility of the model using the proposed method is further demonstrated. Furthermore, the model is validated experimentally using a prototype two-wheeled vehicle.

The vehicle model is used for model-based trajectory planing and is, in this context, compared to a widely used model from the literature.

Since the model is given as a port-Hamiltonian system, a passivity-based trajectory tracking controller is developed using this model. The validity of the controller and the performance of the closed loop are demonstrated using several simulations under different conditions.

Future work

In the future work, the vehicle model may be extended by tire slip models, which leads to relaxing the two nonholonomic constraints to one.

Furthermore, the passivity-based controller is suggested to be validated experimentally. To do so, however, the prototype vehicle needs to be developed further. Using the methods of sensor data fusion as well as global positioning correction data can improvement the vehicle localisation and pose estimation.

Bibliography

[Ada15]   Jürgen Adamy. *Nichtlineare Systeme und Regelungen*. Springer Verlag, 2015.

[AKL05]   K.J. Astrom, R.E. Klein, and a. Lennartsson. Bicycle dynamics and control: adapted bicycles for education and research. *Control Systems, IEEE*, 25(4):26–47, 2005.

[AO03]   A. Astolfi and R. Ortega. Immersion and invariance: a new tool for stabilization and adaptive control of nonlinear systems. *IEEE Transactions on Automatic Control*, 48(4):590–606, April 2003.

[AOAM05]   José Ángel Acosta, Romeo Ortega, Alessandro Astolfi, and Arun D. Mahindrakar. Interconnection and damping assignment passivity-based control of mechanical systems with underactuation degree one. *IEEE Transactions on Automatic Control*, 2005.

[Bet98]   John T. Betts. Survey of Numerical Methods for Trajectory Optimization. *Journal of Guidance, Control, and Dynamics*, 21(2):193–207, 1998.

[Bis16]   Walter   Bislins.   Grundlagen   der   tensorrechnung, http://walter.bislins.ch/physik/media/AR.Chapter2.pdf, March 2016.

[BIW91]   Christopher I. Byrnes, Alberto Isidori, and Jan C. Willems. Passivity, Feedback Equivalence, and the Global Stabilization of Minimum Phase Nonlinear Systems. *IEEE Transactions on Automatic Control*, 1991.

[Blo16]   A. M. Bloch. *Nonholonomic Mechanics and Control*. Number 24 in Interdiciplinary Applied Mathematics. Springer-Verlag New York, Inc., 2 edition, 2016.

[BMCP07]   Pradipta Basu-Mandal, Anindya Chatterjee, and J.M. Papadopoulos. Hands-free circular motions of a benchmark bicycle. *Proceedings of the Royal Society A: Mathematical, Physical and Engineering Sciences*, 463(March):1983–2003, 2007.

[Cos06]   V. Cossalter. *Motorcycle Dynamics*. Gardners Books Limited Eastbourne, 2. english edition, 2006.

[CP10]   Sm Cain and Nc Perkins. Comparison of a bicycle steady-state turning model to experimental data. *Bicycle.Tudelft.Nl*, 2010.

[DBDW05]   Moritz Diehl, Hans Georg Bock, Holger Diedam, and Pierre-Brice Wiebe. *Fast Direct Multiple ShootingAlgorithms for Optimal Robot Control.* Fast Direct Multiple ShootingAlgorithms for Optimal Robot Control. Fast Motions in Biomechanics and Robotics. HAL Id: inria-00390435, 2005.

[DGG⁺17]   O Dong, C Graham, A Grewal, C Parrucci, A Ruina, C Graham, A Grewal, C Parrucci, and A Ruina The. The bricycle : a bicycle in zero gravity can be balanced or steered but not both. *Vehicle System Dynamics, International Journal of Vehicle Mechanics and Mobility*, 3114(August), 2017.

[DP10]   Kees Dullemond and Kasper Peeters. An introduction to tensor calculus, relativity, and cosmology. *Acta Applicandae Mathematicae*, pages 193–, 2010.

[DS12a]   Daniel A Dirksz and Jacquelien M A Scherpen. On Tracking Control of Rigid-Joint Robots With Only Position Measurements B . Tracking Control With Only Position Measurements for. 21(3):1–4, 2012.

[DS12b]   Daniel a. Dirksz and Jacquelien M a Scherpen. Structure preserving adaptive control of port-hamiltonian systems. *IEEE Transactions on Automatic Control*, 57(11):2880–2885, 2012.

[EHP15]   Johannes Edelmann, Martin Haudum, and Manfred Plöchl. Bicycle rider control modelling for path tracking. *IFAC Proceedings Volumes (IFAC-PapersOnline)*, 48(1):55–60, 2015.

[FS99]   Kenji Fujimoto and Toshiharu Sugie. Stabilization of a class of Hamiltonian systems with nonholonomic constraints and its experimental evaluation. *European Control Conference*, 1999.

[FS01a]   K. Fujimoto and T. Sugie. Stabilization of Hamiltonian systems with nonholonomic constraints based on time-varying generalized canonical transformations. *Systems and Control Letters*, 44(4):309–319, 2001.

[FS01b]   Kenji Fujimoto and Toshiharu Sugie. Canonical transformations and stabilization of generalized Hamiltonian systems. *Systems & Control Letters*, 2001.

[FS02]   K. Fujimoto and T. Sugie. Trajectory tracking control of nonholonomic Hamiltonian systems via canonical transformations. *Proceedings of the 2002 American Control Conference (IEEE Cat. No.CH37301)*, 4(July):1–10, 2002.

[FSS04]   Kenji Fujimoto, Kazunori Sakurama, and Toshiharu Sugie. Trajectory Tracking Control of Nonholonomic Hamiltonian Systems via Generalized Canonical Transformations. *European Journal of Control*, 10(5):421–431, 2004.

[FT08]   Kenji Fujimoto and Mitsuru Taniguchi. Passive path following control for port-Hamiltonian systems. *Proceedings of the IEEE Conference on Decision and Control*, (3):1285–1290, 2008.

[GEORA01]   F. Gomez-Estern, R. Ortega, F.R. Rubio, and J. Aracil. Stabilization of a class of underactuated mechanical systems via total energy shaping. In *Decision and Control, 2001. Proceedings of the 40th IEEE Conference on*, volume 2, pages pp.1137–1143, 2001.

[GEvdS04]   F. Gomez-Estern and A.J van der Schaft, A.Schaft. Physical damping in ida-pbc controlled underactuated mechanical systems. *European Journal of Control*, Vol.10(No.5):pp.451–468, 2004.

[GM95]   Neil H Getz and Jerrold E Marsden. Control for an autonomous bicycle. *International Conference on Robotics and Automation*, 1995.

[GMÁF18]   M Ramos García, Daniel A Mántaras, Juan C Álvarez, and David Blanco F. Stabilizing an Urban Semi-Autonomous Bicycle. *IEEE Access*, 6, 2018.

[HPKK14]   Thomas Howard, Mihail Pivtoraiko, Ross A. Knepper, and Alonzo Kelly. Model-predictive motion planning: Several key developments for autonomous mobile robots. *IEEE Robotics and Automation Magazine*, 21(1):64–73, 2014.

[HSF08]   J. Hauser, A. Saccon, and R. Frezza. Achievable motorcycle trajectories. *2004 43rd IEEE Conference on Decision and Control (CDC) (IEEE Cat. No.04CH37601)*, 4:3944–3949 Vol.4, 2008.

[HTI16]   Ryuma Hatano, Takuya Tani, and Masami Iwase. Stability analysis and autonomous stabilization control of a bicycle based on a three-dimensional detailed physical model. *IECON Proceedings (Industrial Electronics Conference)*, pages 324–329, 2016.

[Kel17]   Matthew Kelly. An Introduction to Trajectory Optimization: How to Do Your Own Direct Collocation. *SIAM Review*, 2017.

[Keo08]   Lychek Keo. Trajectory control for an autonomous bicycle with balancer. *2008 IEEE/ASME International Conference on Advanced Intelligent Mechatronics*, 2008.

[KKMN90]   Y Kanayama, Y Kimura, F Miyazaki, and T Noguchi. A stable tracking control method for an autonomous mobile robot. *Proceedings., IEEE International Conference on Robotics and Automation*, pages 384–389, 1990.

[KM97]   Wang Sang Koon and Jerrold E. Marsden. The hamiltonian and lagrangian approaches to the dynamics of nonholonomic systems. *Reports on Mathematical Physics*, 40(1):21–62, 1997.

[KMT09]   Nozomi Katagiri, Yoshitaka Marumo, and Hitoshi Tsunashima. Controller

Design and Evaluation of Lane-Keeping-Assistance System for Motorcycles. *Journal of Mechanical Systems for Transportation and Logistics*, 2(1):43–54, 2009.

[Kot10]   Paul Kotyczka. Transparente Dynamikvorgabe bei der nichtlinearen passivitätsbasierten Zustandsregelung. *TU München*, 2010.

[KS13]   J D G Kooijman and Arend L. Schwab. A review on bicycle and motorcycle rider control with a perspective on handling qualities. *Vehicle System Dynamics*, 51(11):1722–1764, 2013.

[KSM08]   J. D G Kooijman, A. L. Schwab, and J. P. Meijaard. Experimental validation of a model of an uncontrolled bicycle. *Multibody System Dynamics*, 19(1-2):115–132, 2008.

[KVL10]   P. Kotyczka, A. Volf, and B. Lohmann. Passivity based trajectory tracking control with predefined local linear error dynamics. In *American Control Conference*, Baltimore, MD, USA, June 30-July 02 2010.

[KYKY11]   Lychek Keo, Kiyoshi Yoshino, Masahiro Kawaguchi, and Masaki Yamakita. Experimental results for stabilizing of a bicycle with a flywheel balancer. *Proceedings - IEEE International Conference on Robotics and Automation*, pages 6150–6155, 2011.

[Lau98]   J. P Laumond. *Robot Motion Planning and Control*, volume 63. Springer-Verlag Berlin Heidelberg, 1998.

[LLG05]   C. Le Saux, R. I. Leine, and C. Glocker. Dynamics of a rolling disk in the presence of dry friction. *Journal of Nonlinear Science*, 15(1):27–61, 2005.

[LS08]   D. J N Limebeer and Amrit Sharma. The dynamics of the accelerating bicycle. *2008 3rd International Symposium on Communications, Control, and Signal Processing, ISCCSP 2008*, pages 237–242, 2008.

[Lun16]   Jan Lunze. *Regelungstechnik 2*. Springer Vieweg, 2016.

[MAOV06]   Arun D. Mahindrakar, Alessandro Astolfi, Romeo Ortega, and Giuseppe Viola. Further constructive results on interconnection and damping assignment control of mechanical systems: The Acrobot example. *International Journal of Robust and Nonlinear Control*, 16(14):671–685, 2006.

[MP11]   Jp Meijaard and Jm Papadopoulos. Historical Review of Thoughts on Bicycle Self-Stability. *Science Magazine*, 2011.

[MPRS07]   J.P. Meijaard, Jim M. Papadopoulos, Andy Ruina, and A.L. Schwab. Linearized dynamics equations for the balance and steer of a bicycle: a benchmark and review. *Proceedings of the Royal Society A: Mathematical, Physical and Engineering Sciences*, 463(2084):1955–1982, 2007.

[OGC04]   R. Ortega and E. GarcÃa-Canseco. Interconnection and damping assignment passivity-based control: A survey. *European Journal of Control*, Vol.10(No.5):432–450, 2004.

[OS89]   R. Ortega and M. Spong. Adaptive motion control of rigid robots: A tutorial. *Automatica*, Vol.25(No.6):pp.877–888, 1989.

[OSGEB02]   Romeo Ortega, M W Spong, F Gomez-Estern, and G Blankenstein. Stabilization of a class of underactuated mechanical systems via interconnection and damping assignment. 47(8):1218–1233, 2002.

[OVdSMM01] R. Ortega, AJ. Van der Schaft, I Mareels, and B. Maschke. Putting energy back in control. *IEEE Control Systems*, Vol.21(No.2):pp.18–33, Apr 2001.

[OVME02]   Romeo Ortega, Arjan Van der Schaft, Bernhard Maschke, and Gerardo Escobar. Interconnection and damping assignment passivity-based control of port-controlled Hamiltonian systems. *Automatica*, Vol.38(No.4):585 – 596, 2002.

[PCY+16]   Brian Paden, Michal Cap, Sze Zheng Yong, Dmitry Yershov, and Emilio Frazzoli. A Survey of Motion Planning and Control Techniques for Self-Driving Urban Vehicles. *IEEE Transactions on Intelligent Vehicles*, 1(1):33–55, 2016.

[RRPGCG+17] David Rodriguez-Rosa, Ismael Payo-Gutierrez, Fernando Castillo-Garcia, Antonio Gonzalez-Rodriguez, and Sergio Perez-Juarez. Improving Energy Efficiency of an Autonomous Bicycle with Adaptive Controller Design. *Sustainability*, 9(6):866, 2017.

[SAOM13]   I. Sarras, J.Á. Acosta, R. Ortega, and A.D. Mahindrakar. Constructive immersion and invariance stabilization for a class of underactuated mechanical systems. *Automatica*, 49(5):1442–1448, 2013.

[Sch00]   Arjan Schaft. *L2 - Gain and Passivity Techniques in Nonlinear Control*. Communications and Control Engineering. Springer London, London, 2000.

[SHB12]   Alessandro Saccon, John Hauser, and Alessandro Beghi. Trajectory exploration of a rigid motorcycle model. *IEEE Transactions on Control Systems Technology*, 20(2):424–437, 2012.

[SM13]   A. L. Schwab and J. P. Meijaard. A review on bicycle dynamics and rider control A.L. *Vehicle System Dynamics: International Journal of Vehicle Mechanics and Mobility*, (June 2013), 2013.

[SMK12]   A.L. Schwab, J.P. Meijaard, and J.D.G. Kooijman. Lateral dynamics of a bicycle with a passive rider model: stability and controllability. *Vehicle System Dynamics*, 50(8):1209–1224, 2012.

[Sus10]   Leonard   Susskind.   Lessons   on   general   relativity, https://www.youtube.com/playlist?list=PL02EC2EE6A42F70E2,   October 2010.

[TL18a]   Alen Turnwald and Steven Liu. Adaptive Trajectory Tracking for a Planar Two-Wheeled Vehicle with Positive Trail. *IEEE, proceeding*, 2018.

[TL18b]   Alen Turnwald and Steven Liu. A nonlinear bike model for purposes of controller and observer design. *IFAC-PapersOnLine*, 51(2):391 – 396, 2018. 9th Vienna International Conference on Mathematical Modelling.

[TL19]   A. Turnwald and S. Liu. Motion planning and experimental validation for an autonomous bicycle. In *IECON 2019 - 45th Annual Conference of the IEEE Industrial Electronics Society*, volume 1, pages 3287–3292, Oct 2019.

[TO17]   Alen Turnwald and Thimo Oehlschlägel. Passivity-based control of a cryogenic upper stage to minimize fuel sloshing. *Journal of Guidance, Control, and Dynamics*, 40(11):3012–3019, 2017.

[TSL18]   A. Turnwald, M. Schäfer, and S. Liu. Passivity-based trajectory tracking control for an autonomous bicycle. In *IECON 2018 - 44th Annual Conference of the IEEE Industrial Electronics Society*, pages 2607–2612. IEEE, Oct 2018.

[TWB10]   Tae-oh Tak, Jong-sung Won, and Gwang-yeol Baek. Design Sensitivity Analysis of Bicycle Stability and Experimental Validation. *Proceedings, Bicycle and Motorcycle Dynamics 2010*, 2010.

[US06]   N. Uma Shankar. A Fuzzy Controller Design for an Autonomous Bicycle System. *2006 IEEE International Conference on Engineering of Intelligent Systems*, 2006.

[Van05]   Arjan Van Der Schaft. Network Modeling and Control of Physical Systems , DISC Theory of Port-Hamiltonian systems Chapter 2 : Control of Port-Hamiltonian systems. pages 1–36, 2005.

[VB10]   E J H De Vries and J F A Den Brok. Assessing slip of a rolling disc and the implementation of a tyre model in the benchmark bicycle. *Dynamics and Control*, 2010.

[vdSJ14]   Arjan van der Schaft and Dimitri Jeltsema. *Port-Hamiltonian Systems Theory: An Introductory Overview*, volume 1. Foundations and Trends in Systems and Control, 2014.

[VM94]   A. Van Der Schaft and B.M. Maschke. On the Hamiltonian formulation of nonholonomic mechanical systems. *Reports on Mathematical Physics*, 34(2):225–233, 1994.

[VOB+07]   Giuseppe Viola, Romeo Ortega, R. Banavar, José Ángel Acosta, and Alessandro Astolfi. Total energy shaping control of mechanical systems:

Simplifying the matching equations via coordinate changes. *IEEE Transactions on Automatic Control*, 52(6):1093–1099, June 2007.

[Whi99]   F.J.W Whipple. The stability of the motion of a bicycle. *Quart. J. Pure Appl. Math*, 30(120):312–348, 1899.

[WYLZ17]   Pengcheng Wang, Jingang Yi, Tao Liu, and Yizhai Zhang. Trajectory Tracking and Balance Control of an Autonomous Bikebot. *IEEE ICRA, Proceedings*, 2017.

[YCSH14]   Jing Yuan, Huan Chen, Fengchi Sun, and Yalou Huang. Trajectory planning and tracking control for autonomous bicycle robot. *Nonlinear Dynamics*, 78(1):421–431, 2014.

[YKV+14]   Harun Yetkin, Simon Kalouche, Michael Vernier, Gregory Colvin, Keith Redmill, and Umit Ozguner. Gyroscopic stabilization of an unmanned bicycle. *2014 American Control Conference*, pages 4549–4554, 2014.

[Zha14]   Yizhai Zhang. Modeling and control of single-track vehicles: A human-machine-environment interaction perspective. *Dissertation*, 2014.

[ZLYS11]   Yizhai Zhang, Jingliang Li, Jingang Yi, and Dezhen Song. Balance control and analysis of stationary riderless motorcycles. *Proceedings - IEEE International Conference on Robotics and Automation*, 2011.

Curriculum Vitae

Alen Turnwald

Born: 21.09.1983 in Teheran

alen.turnwald@posteo.de

Professional Experiances

Since 03.2019	Elektronische Fahrwerksysteme GmbH (EFS), Gaimersheim
	Development Engineer and member of Audi AI-Lab
11.2014 - 02.2019	Technische Universität Kaiserslautern
	Assistant at Institut for Control Systems
02.2017 - 12.2018	Wilhelm-Büchner-Hochschule, Pfungstadt
	Teacher for control systems and simulation methods
05.2013 - 11.2014	Zentrum für Telematik e.V., Würzburg
	Development engineer in the space technology department
02.2017 - 12.2018	IFW, Leibniz University Hannover
	Student assistant in research and development
02.2017 - 12.2018	Different Institutes of Leibniz University Hannover
	Tutor for Mathematics, Technical Mechanics and Matlab

Education

2011 - 2013	M.Sc. Mechatronics, Leibniz University Hannover
	Robotics and Control Systems
	Overall grade: Excellent (Mit Auszeichnung)
	Thesis: Passivity-based control of a cryogenic upper-stage
2009 - 2011	B.Sc. Mechatronics, Leibniz University Hannover

Award

10.2012	Deutschland Stipendium, Hannover

Ingolstadt, October, 2020

In der Reihe „*Forschungsberichte aus dem Lehrstuhl für Regelungssysteme*",
herausgegeben von Steven Liu, sind bisher erschienen:

1	Daniel Zirkel	Flachheitsbasierter Entwurf von Mehrgrößenregelungen am Beispiel eines Brennstoffzellensystems
		ISBN 978-3-8325-2549-1, 2010, 159 S. 35.00 €
2	Martin Pieschel	Frequenzselektive Aktivfilterung von Stromoberschwingungen mit einer erweiterten modellbasierten Prädiktivregelung
		ISBN 978-3-8325-2765-5, 2010, 160 S. 35.00 €
3	Philipp Münch	Konzeption und Entwurf integrierter Regelungen für Modulare Multilevel Umrichter
		ISBN 978-3-8325-2903-1, 2011, 183 S. 44.00 €
4	Jens Kroneis	Model-based trajectory tracking control of a planar parallel robot with redundancies
		ISBN 978-3-8325-2919-2, 2011, 279 S. 39.50 €
5	Daniel Görges	Optimal Control of Switched Systems with Application to Networked Embedded Control Systems
		ISBN 978-3-8325-3096-9, 2012, 201 S. 36.50 €
6	Christoph Prothmann	Ein Beitrag zur Schädigungsmodellierung von Komponenten im Nutzfahrzeug zur proaktiven Wartung
		ISBN 978-3-8325-3212-3, 2012, 118 S. 33.50 €
7	Guido Flohr	A contribution to model-based fault diagnosis of electro-pneumatic shift actuators in commercial vehicles
		ISBN 978-3-8325-3338-0, 2013, 139 S. 34.00 €

17	Hengyi Wang	Delta-connected Cascaded H-bridge Multilevel Converter as Shunt Active Power Filter	
		ISBN 978-3-8325-5015-8, 2019, 173 S.	38.00 €
18	Sebastian Caba	Energieoptimaler Betrieb gekoppelter Mehrpumpensysteme	
		ISBN 978-3-8325-5079-0, 2020, 141 S.	37.00 €
19	Alen Turnwald	Modelling and Control of an Autonomous Two-Wheeled Vehicle	
		ISBN 978-3-8325-5205-3, 2020, 175 S.	41.00 €

Alle erschienenen Bücher können unter der angegebenen ISBN im Buchhandel oder direkt beim Logos Verlag Berlin (www.logos-verlag.de, Fax: 030 - 42 85 10 92) bestellt werden.